Animals & Men

THE SERPENT DRAGONS OF ANCIENT GREECE

The Cryptozoology of Liberia, a mystery bat from Minnesota, 2015 OOP Butterflies, and much more

The Journal of the Centre for Fortean Zoology   #55

Contents

2. Contents

3. Faculty

4. Editorial

7. Newsfile: New and Rediscovered

14. Newsfile: Chupacabras

15. Newsfile: Blue dogs

16. Newsfile: Man beasts

20. Newsfile: Mystery cats

23. Newsfile: Aquatic monsters

26. CARL MARSHALL'S COLUMN:
The Alcester Seal and other unexpected aquatic guests

33. CARL PORTMAN'S FORTEAN INVERTEBRATES
Gone Batty

37. Watcher of the Skies by Corinna Downes

48. Tasmania Expedition Report by Lars Thomas

56. A little known British pseudocryptid: The Mongoose
by Richard Muirhead

68. Duties for regional representatives—
a discussion document

69. Letters to the Editor

72. Book Reviews

78. Weird Weekend 2015

82. Recent books from CFZ Press

Typeset by Jonathan Downes,
Cover and Layout by SPiderKaT for CFZ Communications
Using Microsoft Word 2000, Microsoft Publisher 2000, Adobe Photoshop CS.
First published in Great Britain by CFZ Press

CFZ Press, Myrtle Cottage, Woolsery, Bideford, North Devon, EX39 5QR

© CFZ MMXV

ISBN: 978-1-909488-41-0

Faculty of the Centre for Fortean Zoology

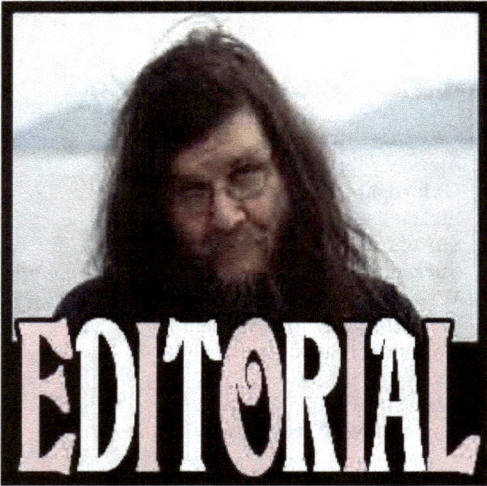

EDITORIAL

Dear Friends,

Welcome to the first issue of this magazine for eleven whole years which does what it is supposed to do.

When I started this magazine back in 1994, it was with promise that I would do my best to make it Quarterly. At first we managed to achieve this goal, but as time went on, our lives became more and more complicated.

The biggest upheaval in our history, came in 2005 when we relocated to North Devon when my father became terminally ill and too frail to look after himself. Upon his death I realised that I had missed living in the country and – quite literally – mortgaged my future by taking out an enormous loan to buy out my brother's share of the old family home. The CFZ trajectory became somewhat diverted by this, and my health problems didn't help. We tried getting involved both with a local tourist attraction and opening something of the kind ourselves, but we found that, not only am I no businessman, but that the global recession of 2008 made such an activity impossible. However, before we discovered this, we lost, and were swindled out of a large amount of money. We also found that as our belts got tighter various people who had been involved with the CFZ had been running me ragged and in some cases robbing me blind, and some people who had been pivotally involved in our activities left suddenly. During all of this time my original vision for an esoteric cryptozoological publishing house became more and more diluted

Sadly, issues of *Animals & Men* became less frequent. In last year's annual report I made a New Year's Resolution in a form of a pledge to the CFZ readership. This was that for the first time since 2003, we would bring the publication schedule for this magazine back to where we wanted it to be in the first place. And I am proud to say, that with the publication of this issue we have done just that.

The Centre for Fortean Zoology has changed massively over the years, both for good and for ill. I would be a liar if I tried to pretend that the last few years have been easy ones. They haven't been. To cut a long story short, the austerity measures of the current British government knocked us for six, and whereas once the CFZ HQ had four or five people working full time, now Graham looks after the animals and the slowly disintegrating buildings, and the administration and publishing (the latter of which I have always considered to be the most important aspect of what we do) is now being run by me, Jessica and Corinna, all – sadly – part-time, because the demands of actually earning a living have forced me and Corinna to take on part-time jobs, and our limited budget means that we can only afford to take on Jessica for 20 hours a week. There are, of course, loyal volunteers

The Great Days of Zoology are not done!

4

around the world who do stuff on an ad hoc basis, but even long term CFZ workers like Richard Freeman have been forced into full-time employment.

However, I am glad to say that we are still – in my humble opinion – the most prolific, and successful cryptozoological research organisation in the world. And I intend to carry on with it for as long as I am able.

To me, one of the most important aspects of what we do; quite possibly *the* most important aspect apart from the publishing is what we do broadly within the theatre of education. For the last ten years or so we have encouraged young people to become involved in what we do, and Jessica is living proof that some of the kids who first came to the Weird Weekend a year ago are even now still involved. Indeed, without the kids who rallied round to help, the Weird Weekend, in particular, would not be able to take place. We have become involved in helping teach the natural sciences to various young people who are being homeschooled, and I feel that this sort of CFZ Outreach is tremendously important if we are to achieve the things I so badly want to do.

The 21st Century is shaping up into a strange and disturbing place. The twin spectres of extreme politics and fundamentalist religion have made this planet a frightening and forbidding place. Things that, even a few years ago, we were able to do with impunity, are now closed to us. For example, it appears that the Indonesian government is now charging potential scientific expeditions into the Kerenci Seblat national park in Sumatra a prohibitive fee which will preclude self-funded organisations like us from going there, sadly, and ironically, the only people who will be able to afford such expensive expedition fees will be the populist television shows who have already done so much to make cryptozoology a laughing stock within the zoological community.

But there is still a lot that we can, and will achieve. Even within the United Kingdom there are many research projects that need to be carried out. And over the next few years we intend to carry out a wide range of these programmes as well as expanding our publications. Here, I would like to thank Dr Andrew May who has worked very hard on dragging our publication website kicking and screaming into the 21st Century where I would have quite happily left it somewhere in the late 1990's. He also finally managed to solve the problems with eBooks which has been tormenting me for many a year. I still do not like them, even though I do end up reading them with monotonous regularity on my trusty iPad. I do accept that electronic publications are a fact of life in these decadent days, and if CFZ publications are to do what I want them to do, then we have no choice but to embrace the new technology.

Because of all of this there has never been a better time to get involved with the CFZ. Whereas once upon a time the CFZ work ethic was geared up purely to people who were able to dedicate relatively large amount of time, now we have restructured so that anyone who can dedicate an hour here or an hour there is gratefully welcomed on board.

These may be strange and disturbing times but they are also fascinating and exciting ones and I implore as many of you as possible to join ranks and help swell the numbers of the CFZ as we approach our quarter century of existence.

Onwards and upwards,

Jon Downes

2015: a Year in the life of the CFZ

Dear Friends,

I can hardly believe that this is the 21st time that I have sat down, a few days before Christmas, and written an account of the previous year's activities for the CFZ members, friends, and supporters.

This is always a bittersweet time for me, because – especially as I grew older – each year brings with it a list of friends, colleagues, and collaborators who have passed on to the Elysian Fields. This year the list is headed by my old friend Daevid Allen who died of cancer in Australia in March at the age of 77. As well as being an important influence on me, musically and culturally, it was through him that I met my long term friend, boss, and CFZ benefactor Rob Ayling.

Long term CFZ cohort Richard Ingram died in the early summer at the age of 68. I had known him for the best part of thirty years, and it is a shock to know that he will never again be a lecturer at the Weird Weekend. Our thoughts are with his son Michael who is left fatherless at an upsettingly young age.

Two important faces from the Bigfoot community died this year. Rob Riggs and Ray Crowe were both affiliated to the CFZ, the former via Nick Redfern, and although I never met either man in the flesh, I carried out a lively correspondence with both for more then a decade. They will be sadly missed.

This has been a complicated, confusing, but ultimately quite a rewarding year. It started with disappointment for Richard Freeman who was struck down by gout the day that he was meant to be flying to Australia for the second CFZ Australia Tasmania expedition, the report of which is in *Animals & Men*, and indeed the expedition report at the 2015 Weird Weekend was given by our old friend Lars Thomas.

The expedition gathered more data to support the idea that the CFZ's totem animal, the thylacine, or Tasmanian wolf, survived its supposed extinction in 1936, and remains a living creature today. It has, after all, been described as 'The liveliest extinct animal alive'.

Arguably more important was the work that Lars Thomas did with the microfauna of the region, obtaining several interesting specimens, some of which appear to be entirely new species to science.

In last year's annual report, I pledged that for the first time since 2003 the CFZ flagship journal *Animals & Men* would meet its stated publication schedule of four issues a year.

We bid Farewell to Richard and Daevid

In order to do this we underwent the biggest shake up of our publication schedule, and – indeed - our membership structure for the past twenty-one years. For the first time ever, we removed the subscription model, and *Animals & Men* began to be published in three different formats.

1.   An online flipmag embedded on the CFZ publications website and reachable through links on the blogs and main site. This is, and will remain free.

2.   A hard copy, traditional magazine available at low cost from Amazon and all associated outlets.

A Kindle edition formatted to be read on electronic book readers.

In January, we were approached by our old friend and colleague Dr Andrew May who offered to sort out the increasingly outdated CFZ publications website. This turned out to be a massive undertaking, and by the time he and I had finished, the main CFZ website had been given a stringent, and brand new CFZ publications website and the CFZ shop had been built from scratch.

For some years we have been trying to arrange publication of our books and magazines in eBook format. For a number of reasons this never happened, causing a great deal of conflict within the CFZ community. Andrew May sorted it out once and for all, and now, something in the region of a third of our publications are available in eBook format. As anybody who has ever dipped their toe into the increasingly murky waters of internet publishing will know, there are quite a few different formats that eBooks can be published in, most of them are incompatible with each other.

We decided to publish in the Kindle format

because there are three apps available in most other formats which mean that Kindle books can be read on Android, iOS, and PC formats (amongst others).

We would like to publicly thank Andrew May for all of his hard work this year and for very patiently explaining the intricacies of internet publishing to an old duffer like me. I truly do not know what we would have done without him.

Another important change took place this summer. Because of an increasing work load, and – I have to admit – my failing health, I have been getting further and further behind with the things that I have to do. Regular CFZ watchers will know my adopted niece Jessica Taylor; she has been doing voluntary work with us on an ad hoc basis since she was about 12 and hanging out in my office playing *Zoo Tycoon* for some years before that. When I decided that I had to bite the bullet and engage a proper assistant for 20hours a week, Jessica, fresh from studying Business Studies at the local college, was the only realistic candidate. She started working for us in July and has been bullying me (although she insists that it is for my own good) ever since.

Joking aside she is a lovely girl and I always look forward to the days that she is in the office.

The previous CFZ interns, Saskia England and Sheri Myler have both left. Sheri has now graduated and is working in a national park in the North of England, and Saskia is studying marine biology in Plymouth University. Therefore there were two large intern-shaped holes in the CFZ infrastructure. We are very happy to say that Charlotte Phillipson and Nadine Rider have joined the team as interns. I am giving both of them lessons in natural history, officially to Nadine because she is

home schooled. I am very much an old school naturalist and am appalled at the way the modern education system seems to roundly ignore most of the things I consider to be important., so in my own little way I am trying to redress the balance.

This year we also took on another one of my adopted nephews, Danny Owens, to replace Mark Raines as gardener and groundsman after Mark left the village. Seldom have we had anyone as hard working and diligent as Danny, and I only wish we could afford to take him on full time.

The Weird Weekend was held on the third weekend of August, once again at The Small School. This year's speakers were:

- Nick Wadham: Wild and Deadly Animal Show
- Lee Walker: Urban Legends of Liverpool
- Lars Thomas: Microcyrptozoology
- Judge Smith: Ouija Boards
- Shoshannah Hughes: Feral Cats
- Rob Cornes: The Seal Serpent:
- Ronan Coghlan: Irish Cryptozoology
- Rosie Curtis: Scary Internet Memes
- Steve Rider: Tales from the Infinite
- Jaki Windmill: Astroshamanics
- Richard Freeman: Dragons
- Adam Davis: Manbeasts and Me
- Lars Thomas: Tasmania expedition Report
- Richard Muirhead: Mystery Animals of Hong Kong

One of the highlights of the weekend came when Carl Marshall, Lars Thomas and I presented the latest evidence in our long standing investigation into the possibility of a hitherto unrecorded mammal species for the UK, the beech marten. Much to my surprise and pleasure, Carl presented me with a taxiderm specimen of one of these creatures believed to have been taken in Dorset during the 19th Century; seldom have I had a better birthday present.

We continued with our programme of publishing this year, with the following titles:

- More Stars Steeds and Other Dreams: The Collected Poems. By Dr Karl P N Shuker
- Sasquatch Down. By Michael Newton
- The Song of Panne (Being Mainly about Elephants). By Jonathan Downes
- Weird Wessex: A Tourist Guide to 100 Strange and Unusual Sights. By Andrew May and Paul Jackson
- Glimpses in the Twilight. By Lee Walker
- Brundannon's Daughter; Through the Realms of the Woodwose. By Corinna Downes
- Going Mad to Stay Sane. By Andy White
- In the Footsteps of the Russian Snowman. By Dmitri Bayanov
- The Scribbling Sea Serpent. By Kate Kelly
- Mysterious Creatures: A Guide to Cryptozoology – Volume 2. By George Eberhart
- Strange Skies, Strange Eyes. By Brian Allan

We have also published volume three of the *Journal of Cryptozoology*, edited by my old friend Dr Karl Shuker and we look forward to publishing volume four in 2016.

The 2016 publishing schedule includes major new books by: Dr Karl Shuker, Richard Muirhead, S .D. Tucker, Rob Cornes, Brian Allan, Matt Salusbury, Richard Freeman, and more.

As mentioned earlier, this year saw a major reorganisation of the CFZ membership package. As a result of this, in the autumn we instituted a new CFZ monthly members' newsletter, the third issue of which will be emailed out on the 1st of January. Reaction to this has been extremely favourable so far, and we look forward to it becoming a much loved and welcome edition to the CFZ publication schedule.

2016 will see the return of the CFZ Yearbook, and – after all of the upheavals of recent years - we hope that it will be back as a regular event in the CFZ publishing schedule. Something else that has been missing in recent years has been our monthly web TV show and we are pleased to announced that plans are underway to bring it back in earlier 2016, co-presented by Charlotte Phillipson.

We have this year had funding difficulties, entirely related to the loss of income from my house in Exeter and the length of time it has taken to effect the repairs necessary. This has severely affected our cash flow, and is purely the result of us trusting the wrong people. And to ensure you that things will return to normal soon, it only remains for me to wish you all a Happy Christmas and a peaceful and prosperous 2016.

Jon Downes
(Director CFZ)

Newsfile

The Whistleblowing Cray

Edward Snowden is best known as the man who blew the whistle on the US National Security Agency's surveillance activity, but thanks to researchers at the Humboldt University of Berlin, he now has another claim to fame – as the namesake for a new species of crayfish. The new species, which is described in a recent edition of the journal *ZooKeys*, has been named the *Cherax snowden* and was found in the freshwater tributary creeks in West Papau, Indonesia, by German scientist Christian Lukhaup and his colleagues, *The Washington Post* reported.

So what made him Lukhaup name this new creature after Snowden, who leaked top-secret NSA documents to a trio of journalists back in 2013, exposing the agency's surveillance program? He wrote that it was because he viewed the controversial figure as an "American freedom fighter".

"After describing a couple new species, I thought about naming one after Edward Snowden because he really impressed me," he told the newspaper. "We have so many species named after other famous people who probably don't do so much for humanity. I wanted to show support for Edward Snowden. I think what he did is something very special."

SOURCE: http://www.redorbit.com/news/ science/1113408210/new-species-of-crayfish-named-in-honor-of-edward-snowden-082715/ #MFOq0diZL1Rd1EWe.99

Sounds a bit Harry Potterish

Ophidiophobics should fret not, but Australia has a new species of snake. Scientists have identified a new type of death adder in the Kimberley region of Western Australia.

Named the Kimberley death adder, or _Acanthophis cryptamydros_, the snake is about 50cm long and has a diamond-shaped head. Scientists from Australia and the UK discovered the snake is different from the death adders found around Darwin in the Northern Territory. Previously it was thought the same species extended to the Kimberley, but an examination of 20 snakes found in the WA region has its very own death adder.

The death adder family is, true to their moniker, considered to be among the most venomous snakes in the world. Before antivenom was available, around half of the people bitten by death adders suffered paralysis and death.

SOURCE: http://www.theguardian.com/environment/2015/sep/16/yet-another-deadly-snake-species-discovered-in-australia

U dig that?

After being mostly neglected for more than a hundred years, a group of digger wasps from Australia has been given a major overhaul in terms of species descriptions and identification methods.

This approach has led to an almost 50% rise in the number of recognized species of these wasps on the continent. The study was published in the open access journal *ZooKeys*.

They call them with names like "Great Golden Digger" or "Great Black Wasp" in the US and there is a good reason behind it. However, some of these digger wasp species do not impress solely with their looks, but also with their wide range of distribution. Members of the wasp genus Sphex can be found in almost every area of the world.

Two researchers from the Museum für Naturkunde in Berlin, Thorleif Dörfel and Dr. Michael Ohl have now reexamined the species diversity of Sphex in Australia. They have identified eleven new species.

SOURCE: http://www.sciencedaily.com/releases/2015/09/150917135007.htm

The Giant Eat of Sulawesi

Scientists have discovered a new species of rat in Indonesia with a strange mix of features never seen before, including a hog-like snout, big ears, a tiny mouth and "curiously" long pubic hair. The rat, first discovered by scientists in 2013 in the remote mountain jungles on the island of Sulawesi, has been named *Hyorhinomys stuempkei,* or hog-nosed rat. Scientists said the creature was so unique it has been listed as an entirely new genus. "I am still amazed that we can walk into a forest and find a new species of mammal that is so obviously different from any species, or even genus, that has ever been documented by science," said Dr Kevin Rowe, a researcher from Museum Victoria in Australia.

SOURCE: http://www.telegraph.co.uk/news/worldnews/asia/indonesia/11913909/Bizarre-hog-nosed-rat-discovered-in-Indonesian-jungle.html

CLAWED FROG EXPANSION

Researchers have discovered half a dozen new species of the African clawed frog, and added back another to the list of known species, in the process uncovering striking new characteristics of one of the most widely studied amphibians in the world. The discovery increases the number of known clawed frog species from 22 to 29 -- a 30 percent increase. "Because the African clawed frog is used as a model organism for biological research, it would be understandable to think that scientists had already pinned down the number of species and other aspects of their diversity such as where they live and how they are related to one another," says Ben Evans, lead author of the study published in *PLOS ONE* and an associate professor in the Department of Biology at McMaster University. "But this isn't the case."

These clawed frogs, found in west and central sub-Saharan Africa, live in slow moving or stagnant water and are characterized by flattened bodies, vocal organs which can produce sound underwater, and claws on its first three toes. Researchers were able to tease apart a wealth of information on its evolution with new analytical techniques using DNA, voice recordings, CT scanning of internal anatomy, chromosome analysis and more.

SOURCE: http://www.sciencedaily.com/
releases/2015/12/151216151615.htm

FROG PITCHES AND CATCHES

Two new arboreal frog species have been discovered in New Guinea, researchers have revealed in a recent study. Both are considered the first members of the frog genus Cophixalus found on Misool Island. One of the frogs was also identified as a hermaphrodite, an animal that has both male and female reproductive organs. Misool Island is one of the four major islands in the Raja Ampat Islands in the Indonesian part of New Guinea. After heavy rainfall swept through the islands one night, researchers from the South Australian Museum set out to track and record frog mating calls. They also collected tissue samples for DNA analysis and took photographs. It wasn't long before they realized they had located two new Cophixalus frogs among the logged lowland rainforests. All the specimens collected have since been placed in Indonesia's Museum Zoologicum Bogoriense. Researchers on the nearby islands of Batanta and Waigeo failed to find these new species despite similar climatic condition and strong frog activity, which, researchers says, only underscores the vast biodiversity present among the Raja Ampat Islands. The findings were recently published in the journal *Zoosystematics and Evolution*.

SOURCE:http://www.natureworldnews.com/
articles/17261/20151007/two-new-frog-species-found-guinea.htm

Another Compendium of Frogs

CAUGHT IN CUSCO

A new species of frog was recently discovered in Cusco, reports the National Service of Protected Areas by the State (Sernanp). The new species was found in the National Sanctuary Megantoni in the protected natural area of Cusco.

The terrestrial frog species was discovered by park rangers and protected area specialists at the converging point of the Andes and Amazonian slopes between 3,506 meters and 3,651 metres above sea level. The new species is said to belong to the genus Bryophryne, and is unique for its smooth dorsal skin with scattered olive green warts. In addition, it lacks the basal membrane and keels in its fingers. The frog was found at the limit of the Sanctuary and near Manu National Park. Specialists and researchers from the Museo de Historia Natural de la Universidad Nacional San Agustin de Arequipa were conducting a routine biological study when they found the new species.

SOURCE: http://www.peruthisweek.com/news-new-frog-species-discovered-in-cusco-107813

GHATS ENTERTAINMENT

A new species of tree frog was discovered and another species of bush frog was rediscovered by a team of researchers during their recent exploration in the Western Ghats.

The discovery, published in the latest issue of International Taxonomic Journal *Zootaxa*, is a joint effort by a team comprised Robin Abraham, a researcher from the University of Kansas, USA; Anil Zachariah, a batrachologist from Wayanad and Vivek Philip Cyriac, a researcher, of the Indian Institute of Science Education and Research, Thiruvananthapuram. The team discovered a new species of Rhacophorid tree frog of the

genus Ghatixalus. The new species is named *Ghatixalus magnus* after its large size making it the biggest known tree frog from the Western Ghats.

The team has also rediscovered a frog that had been evading for the past many decades. *Raorchestes flaviventris*, a species of rhacophorid bush frog described from the Western Ghats by George Albert Boulenger in 1882, had never been reported from the region since its description. The team members have found this elusive frog from the High Ranges of Idukki district in Kerala. The discovery of this frog after around 132 years was published in International Journal *Zootaxa*.

SOURCE: http://www.thehindu.com/todays-paper/tp-national/new-species-of-tree-frog-discovered/article7913832.ece

Another Compendium of Frogs

But it doesn't eat fish

What Darwin Taught Us

Three new species of fishing snake have been identified as a result of both field and laboratory work, undertaken by Dr. Omar Torres-Carvajal, Museo de Zoología QCAZ, Ecuador, in collaboration with herpetologists from Peru (CORBIDI) and the United States (Francis Marion University). The new species differ from their closest relatives in scale features, male sexual organs and DNA. The unusual discoveries took place in areas within the 1,542,644 km2 of the Tropical Andes hotspot, western South America. Although they are commonly known as fishing snakes, these reptiles most likely do not eat fish. Their diet and behaviour are poorly known. So far, it has only been reported that one species feeds on lizards.

The fishing snakes have long been known to live in cloud forests on both sides of the Andes of Colombia and Ecuador. Yet, it seems they have waited all along to make an appearance. The new species described herein, along with a recent description of one species from southwestern Ecuador also published in _Zookeys_, has duplicated the number of species of fishing snakes from four to eight over the span of several months.

SOURCE: http://www.sciencedaily.com/releases/2015/12/151216134421.htm

Paging Charles Darwin: The island of Santa Cruz within the Galápagos has not one but two distinct species of giant tortoise, a new genetic study finds.

For years, researchers thought that the giant tortoises living on the western and eastern sides of Santa Cruz belonged to the same species. But the tortoises look slightly different, and so recently scientists ran genetic tests on about 100 tortoises from both groups.

The tests were definitive: The two tortoise populations, which live only about 6 miles (10 kilometers) apart on the opposite sides of the island, are actually extremely distant relatives.

The Santa Cruz tortoise species that has long been called _Chelonoidis porter_ are the ones living on the western side, in a region of the island known as La Reserva. And now, the newly identified eastern Santa Cruz tortoise has been named _Chelonoidis donfaustoi_. It inhabits an area known as Cerro Fatal.

SOURCE: http://www.livescience.com/52545-new-species-galapagos-tortoise.html

Chupacabras

Another Dead Monkey

Once again the decomposing corpse of a perfectly well known species has been found in a river. And once again the global media jump on it as proof of the legendary chupacabras. This bloated corpse washed up in Paraguay in mid-October and for a few days was causing a sensation in the local media, with some speculating that the body is that of the legendary chupacabras. However, experts believe the creature pulled out of a river at Carmen del Parana is something a lot more common -- and a lot less sensational.

Javier Medina, head of the local fire department, told 780 AM that it has "little hands and feet," leading him to believe it's a monkey, but the body is so badly decomposed it's hard to say for sure. Of course it is a monkey. So-called cryptozoologists who make a career out of such things should be ashamed of themselves.

SOURCE: http://www.huffingtonpost.com/entry/chupacabra-paraguay_562eccbfe4b00aa54a4af6c9

Blue Dogs

A series of trail cam pictures from the vicinity of Topeka, Kansas, have led to various newspaper pundits proclaiming the existence of a new cryptid - the Kansas Chupacabras. Well, whatever they are, they are definitely not chupacabras, as we have explained *ad nauseam* in these pages over the years.

However, one of the pictures, an extract from which we have inset here, does appear to show the strange pads which appear on the haunches of the Texas blue dogs. What does this prove? Nothing at the moment, but the case is going into our burgeoning case files.

SOURCE: http://tinyurl.com/pqz8saz

Man Beasts (BHM)

Up the Provo

The following report comes from Provo Canyon, Utah this summer, although it has been suggested that the event took place as long as three years before.

"We went camping in Provo Canyon (near Squaw Peak and Little Rock Canyon Overlook) and saw some deer up on a hill that we wanted to get a closer look at. On our way up, we thought we saw a bear, until the monster stood up and looked right at us. We ran straight to the car after that, leaving our tent and everything behind. It's probably all still up there."

SOURCE: http://consciouslyenlightened.com/video-caught-on-film-bigfoot-sighting-in-provo-canyon/

Wisconsin Waderer

This story comes from the 'Bigfoot Evidence' blog, which - increasingly - is becoming one

of the main sources for bigfoot stories. It was posted in mid-December and reads: "A couple were fishing in the Forks, WA area inside the Hoh Rainforest when they noticed these large tracks on the river trail. They aren't sure what made the tracks, but they are quite large, and barefoot. They also mentioned that it was currently 38 degrees. It's difficult to believe someone would be walking around barefoot when it's that cold, even if they did have giant feet."

SOURCE: http://bigfootevidence.blogspot.co.uk/2015/12/couple-finds-huge-footprints-in-hoh.html#moretop

Why did the Sasquatch cross the road?

Another one from Bigfoot Evidence:

"ECBRO investigator Daniel Benoit caught some very fresh humanoid tracks crossing a remote road headed into the woods. So fresh they were still wet! Whatever crossed the road had human-like feet, and did it right before he got there. Why would a human be out walking through the woods in the middle of nowhere, barefoot in December? Doesn't make much sense."

SOURCE: http:// bigfootevidence.blogspot.co.uk/search? updated-max=2015-12-11T14:00:00- 08:00&max-results=20&start=60&by- date=false

The Truth is In There

It is always interesting when official documents emerge indicating that those in power over us are at least prepared to admit the possibility that cryptids exist in physical form. In the last series (or maybe the one before) of the hit American TV series 'Bones', one of the characters claimed that whilst with a Special Forces team in Pakistan he was a witness to the yeti. Now official documentation has emerged from the 1950s suggesting that the American Government was prepared to admit that this might actually happen.

In the lead up to Halloween, the U.S. National Archives showcased some of its "creepiest" documents. One of the winners? The official policy from the State Department telling travellers what to do in the event that they encounter a Yeti while exploring Nepal. There is always the possibility, I suppose, that this was put there as an amusing hoax to amuse the public, but it seems too much for them to be bothered with.

The news story that we received reads:

NATIONAL ARCHIVES

"Highlights include the instruction that the Yeti "must not be killed or shot at except in an emergency arising out of self-defense," as well as the requirement that if you see a Yeti, you should tell the Nepalese government first, and not some journalist. As if.

In the 1950s, when cryptozoology was emerging, there were several publicized expeditions to Nepal by wealthy Americans who wanted to prove the existence of Yetis. The U.S. government has a form for everything, so it makes sense that Foreign Service workers would churn out some guidelines about a mythical beast.

Seeing as every use of the word "Yeti" is in quotation marks, it's probably safe to assume that Ernest H. Fisk did not believe in Abominable Snowmen and would have rather spent his time issuing memos about literally anything else. Then again, in 1959, you probably had to make your own entertainment at the U.S. embassy in Katmandu."

SOURCE: http://fusion.net/story/221029/ yeti-us-department-state-nepal/? utm_source=facebook&utm_medium=social &utm_campaign=atlas

AIR POUCH
PRIORITY

UNCLASSIFIED
(Security Classification)

FOREIGN SERVICE DESPATCH

DO NOT TYPE IN THIS SPACE

031.0090c/11-3059

DEC 10 1959

FROM : American Embassy, Kathmandu

DESP. NO. 73

TO : THE DEPARTMENT OF STATE, WASHINGTON.

DATE: November 30, 1959

REF : Embassy, New Delhi, Despatches 1473, June 22, 1959 and 374, September 30, 1958

For Dept. Use Only	ACTION	DEPT.
	NEA-4	IN: RM/R-1, IRC-8, L-2, M/O-1, M/OP-1, IES-5, ICA-11, S/SA-1
	REC'D 12/7	OTHER: CIA-10, USIA-10, OSD-4, OCB-1, ARMY-4, NAVY-3, AIR-1

SUBJECT: REGULATIONS GOVERNING MOUNTAIN CLIMBING EXPEDITIONS IN NEPAL - RELATING TO YETI

There are, at present, three regulations applicable only to expeditions searching for the YETI in Nepal. These regulations are to be observed in addition to the 15 clauses as listed in Mountaineering and Scientific Expeditions in Nepal.

The three regulations are as follows:

1. Royalty of Rs. 5000/- Indian Currency will have to be paid to His Majesty's Government of Nepal for a permit to carry out an expedition in search of "Yeti".

2. In case "Yeti" is traced it can be photographed or caught alive but it must not be killed or shot at except in an emergency arising out of self defence. All photographs taken of the animal, the creature itself if captured alive or dead, must be surrendered to the Government of Nepal at the earliest time.

3. News and reports throwing light on the actual existence of the creature must be submitted to the Government of Nepal as soon as they are available and must not in any way be given out to the Press or Reporters for publicity without the permission of the Government of Nepal.

FOR THE AMBASSADOR:

Ernest H. Fisk
Counselor of Embassy

JROLingermans:ml
REPORTER

UNCLASSIFIED

INFORMATION COPY

Mystery Cats

The Great Eastern

I have noted before the slightly distasteful way that the American government has treated the reappearance of pumas in the Eastern and Central States of the USA. It all gets a little complicated.

Historically it was believed that, apart from several subspecies in Western North American, there were two in the East of the country:

1.   The Eastern cougar
2.   The Florida panther

As most readers will know the Florida panther has become extremely endangered, and the Eastern puma was declared extinct in June 2015, after many years of no confirmed reports. However, this is not to say that there weren't any reports in those years, and many cryptozoologists believed that the Eastern puma was not extinct.

Well, despite a plethora of sightings, photographs, and other evidence the animal certainly is extinct now, in a very real sense! Because it was not long after the subspecies was declared extinct that a speculative paper from 2000 suggesting that all pumas in North America were all of the same species (admittedly something which the legendary John Audubon believed as long ago as 1851) has been accepted as canon.

The pumas which are turning up across Central and Eastern USA are therefore being designated as escaped pets, and stragglers from Western populations. To those of us of a suspicious frame of mind might conclude that the American government has a lot to gain from this. They no longer have to give the two subspecies expensive protection, and they may even be able to be shot with impunity. But we all know that politicians have everyone's best

06:35PM 11/24/2015

intentions at heart, don't we?

In late November, a trail camera caught sight of a cougar in Humphreys County, Tennessee, 70 miles west of Nashville, News Channel 5 reports. There have been other big-cat sightings in West and Middle Tennessee this year that the Tennessee Wildlife Resources Agency has verified, as well. TWRA spokesman Don King says that a cougar "[had] not been confirmed by state authorities in Tennessee in perhaps 100 years" until these recent reports.

Speculation about Tennessee mountain lions has been a source of myth and lore in the region for about as long. Historically, locals referred to the cats as "panthers," and their existence in the past is well known. According to the Center for Biological Diversity, the last indigenous Tennessee cougar was killed in 1930, and biologists assert that the cats responsible for the recent spottings have migrated from the West, through the Midwest, and now into the Southeast.

SOURCE: http://www.fieldandstream.com/blogs/field-notes/tennessee-officials-confirm-first-cougar-sighting-in-100-years

THE Plymouth Panther

Office worker Carole Desforges spotted a mystery animal prowling on the lawn opposite her home on the outskirts of Plymouth in Devon.

She managed to get a few snaps of it through the window before it ran off. Carole, 59, first thought it was just a fox - but having reviewed her pictures she is now not so sure.

SOURCE: http://www.mirror.co.uk/news/uk-news/big-cat-sightings-huge-black-5990677

Aquatic Monsters

Only Ogopogo

"It is true, it does exist, I saw it with my own eyes," says a Kelowna boater of a surprise experience with Ogopogo. Two women enjoying the sun and some fun time on the water had the scare of their life when they say Ogopogo swam near their boat.

"It was a huge snake, I saw it, I saw the head. It was two feet thick and it was like 50-feet long. I could not believe it," says Suzie St-Cyr Cowley. The "Ogopogo" was spotted about 1,000 feet off shore in front of Quails Gate Winery on the westside. She says it was heading south in the lake. "I could not believe it," says St-Cyr Cowley. "I was afraid because we were so close and I wanted to move my boat away. I was screaming "Oh my god, that's Ogopogo! It was so big."

SOURCE: http://www.castanet.net/news/Kelowna/138167/Boaters-say-they-saw-Ogie

My Cup of Tay

Retired Council Chief, Provost and Lord Lieutenant Mervyn Rolfe claims he saw a monster swimming in the River Tay near Dundee. He took a photo showing three black humps looming out of the water. The story as printed in *The Daily Morror* was even more scanty than usual and claims that it was Nessie on 'her' summer holidays. He also said how the object was followed by dolphins. Reading through the accounts of this story in half a dozen different newspapers it is impossible to glean any more information, and we suspect that it is probably driftwood. However, we are sure that all readers of this magazine are aware of our collective respect for authority figures, especially politicians, and so we would not dream of contradicting the claim that this is one of Nessie's holiday snaps,

SOURCE: http://www.mirror.co.uk/news/uk-news/loch-ness-monster-enjoying-summer-6365158

Arma and the Man

Inverness cruise company Loch Ness by Jacobite wants to replace the unicorn, a legacy from William I's decision to use the mythical creature on his coat of arms. The first sighting of the Loch Ness Monster dates back to 565AD and the question of whether or not she exists is said to be worth millions each year to Scottish tourism. Freda Newton, of Loch Ness by Jacobite, said: 'We have been running tours of Loch Ness for 40 years now, with many of our visitors coming to search for or at least catch a glimpse of one of the world's most famous monsters.

'Nessie is an icon and an asset. There is no doubt she attracts hundreds of tourists to Scotland every year and she deserves recognition. If not as our national animal, then at least she should be awarded the title of Scotland's national monster. So a campaign has been launched to have Nessie recognised as the national animal of Scotland.

SOURCE: http://metro.co.uk/2015/04/23/loch-ness-monster-could-become-national-animal-of-scotland-5162923/

The Eel Truth

A conger eel said to be up to 21ft in length when it was caught off the coast of Devon was actually only about 6-7ft, officials at the port where it was sold have said. The fish weighing 72.5kg (160lb) was hauled aboard the inshore trawler Hope by fishermen from Plymouth who tweeted pictures of their catch. According to Plymouth Fisheries the eel is thought to be around 6-7ft (2m) long, though the fisherman who hauled it onboard said it could have been a few feet longer. A spokeswoman for Plymouth Fisheries said: "Sadly it was never measured before it went to auction but the fisherman who caught it estimated it was 'about 10 feet in length'. "Possibly the perspective of the hanging photograph does make it look longer - it looks shorter in other photos and eels stretch out when hung."

SOURCE: http://news.sky.com/story/1484518/the-eel-story-giant-fish-big-but-not-that-big

Monster Mashu

MONSTER MASHU

The clearest lake in the world is Lake Mashu, located in the northeastern part of Hokkaido, Japan. Known to the native people, the Ainu, as "Kamuytou" or "The Lake of the Gods, it's a caldera lake, formed in the crater of a dormant volcano approximately 11,000 years ago. Lake Mashu is notable for having the clearest water in the world, which made it easy for a team of researchers to check out the bottom of the lake. The research team from Japan's National Institute of Environmental Studies discovered strange animal track marks in the bottom of the lake. The trail resembles marks left by a tank. Or perhaps a crayfish? Since the 1970's there have been reports of giant crayfish living in Lake Mashu. Some claim sightings of crayfish far exceeding the size of any known to be in Japan.

SOURCE: http://tinyurl.com/phkc4k3

We mean it Ma'am

It is a mystery which has baffled and enthralled generations, sparking scientific studies, countless theories and repeated searches of the murky deep.

Now it can be revealed that such was the public fascination with the Loch Ness Monster that the Queen was at one point asked to agree to the creature being named after her. Newly discovered papers show that Sir Peter Scott, the eminent conservationist who led the search for the monster in the early-sixties, wrote to the palace with the suggestion in 1960.

He proposed that should the Loch Ness Monster – or Nessie as it was affectionately known – be eventually found it could be named 'Elizabethia nessiae'.

Palace officials poured cold water on the idea, fearing that the Queen might become associated with what could eventually turn out to be an embarrassing hoax.

SOURCE: http://www.telegraph.co.uk/news/uknews/queen-elizabeth-II/11846063/Revealed-The-plan-to-name-the-Loch-Ness-

Odds and Sods

Harvey Hops

A series of stories about "giant jackrabbits" which are described as being "the size of dogs" have been reported from Fargo, North Dakota. However the description is a particularly useless one as that could mean anything between a Chihuahua and a Great Dane. Kayla Straabe told ABC News in February.

"Every day, I feel like the crazy rabbit lady chasing them out of the yard where they're having a hay day," Straabe said. "There's at least 40 to 50 everyday, and they're in our yards and by a children's park."

She said the city pest control department told her that they couldn't do anything about the jackrabbits, technically wild hares, because they were wild animals "I was told we should poison them, which I will absolutely not do," Straabe said.

There is no city ordinance that deals with depopulating rabbits in the area, Fargo

Jack Rabbits Invade Neighborhood

Police Lt. Joel Vettel told ABC News. "I can't imagine someone would suggest poisoning the rabbits, but it's true we don't have anything in place to allow us to deal with the rabbits," Vettel said. "What residents can do is start a formal process to get an ordinance, which is usually done at committee meetings."

This tells us more about local politics than it does lagomorph gigantism, but we await further details with interest.

SOURCE: http://abcnews.go.com/US/dozens-dog-sized-jackrabbits-development-north-dakota/story?id=28662261

Teratology

Neon Jackfish

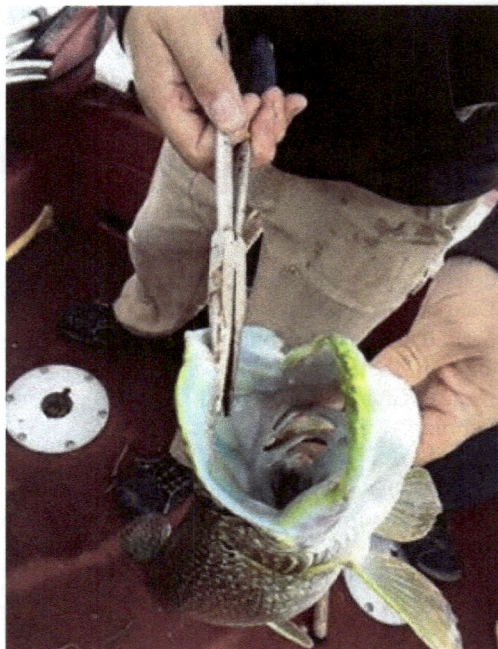

It's not the Loch Ness monster, but a Yellowknife angler has ignited debate of his own after landing, and then releasing, a fluorescent green pike while fishing in Great Slave Lake. The pike's lower lip was the most discoloured part, said Straker. 'It looked like green lipstick. It was so bright.' "The whole top of the fish had a different green," said Randy Straker. "If you look at the mouth, it looked like green lipstick. It was so bright."

Straker was fishing with his friend Craig Thomas on Sunday in the lake's North Arm when he made the catch. After pulling the pike into the boat — Straker estimated it at 38 to 40 inches and 12 to 14 pounds — the two men snapped a few photos and released their catch. Afterward, though, they realized that they'd caught something quite unique.

SOURCE: http://www.cbc.ca/news/canada/ north/neon-jackfish-leaves-yellowknife-fisherman-stumped-1.3204762

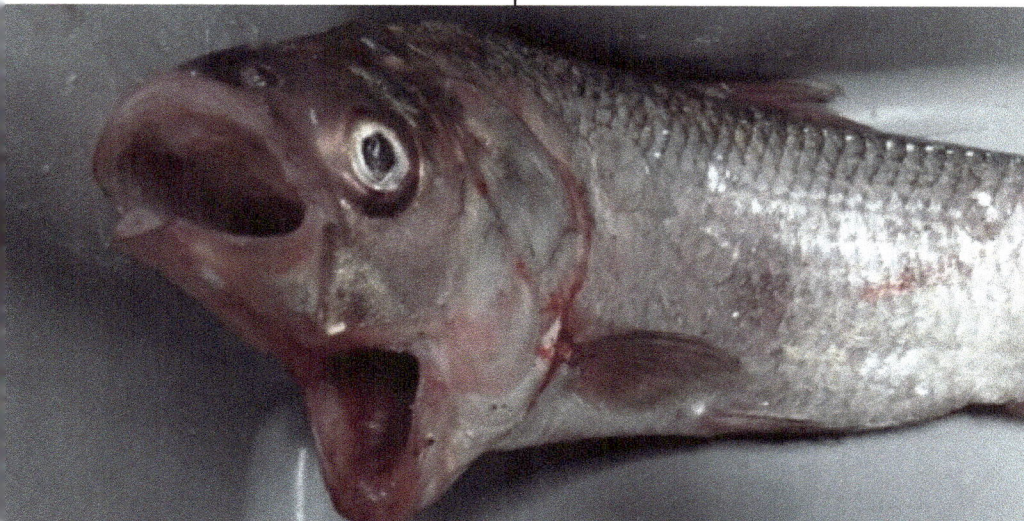

Two mouths to bite you

As if to warn us not to eat too much seafood a fisherman in Australia caught a fish with two mouths. What's in the water down there?

The fish is a bony bream (*Nematalosa erebi*), a freshwater fish that can reach 32 cm (12.6 inches) in length and goes by a variety of interesting names such as Australian River Gizzard Shad, Hairback Herring and Melon Fish. You can add the Double Mouth Lake Bonney Monster to the list as this one was caught in the coastal lake located in the south east of South Australia. What caused this bony bream to end up with two mouths? It could be chance. It could be because Australia is famous for strange fish, as seen recently with the rare catches of a goblin shark and a frilled shark. It could be because Lake Bonney has been getting wastewater from nearby pulp and paper mills for over 60 years.

Animal oddities are always a cause for concern, especially as their numbers continue to rise. That's the real warning. Let's just hope this one isn't a sign that fish are evolving into politicians.

SOURCE: http://mysteriousuniverse.org/2015/02/fish-with-two-mouths-may-be-a-warning/

At night all foxes are...

At the end of September a string of newspaper articles proclaimed that Britain's rarest animal had been spotted. How they figure that a melanistic colour variant of the red fox is 'Britain's rarest animal' the A&M editorial team are not too sure, and would probably dispute the terminology if we could be bothered.

The story reads:

"The rarest animal in Britain has been spotted in none other than North Yorkshire, after one man caught a black fox on camera.

Robert Fuller was notified by one of his neighbours that a fox had been prowling around his property, and upon taking a closer look, he realised it was a black fox – the rarest animal in Britain.

The wildlife photographer from North Yorkshire recognised the fox immediately due to his line of work, and began taking pictures of the creature, to document seeing it."

We would certainly dispute that it is the rarest animal in Britain—that honour probably goes to the pool frog, but it is certainly a striking beast!

SOURCE: http://www.unilad.co.uk/articles/britains-rarest-animal-has-been-spotted-in-yorkshire/

As you may have noticed, there are considerably more news items this issue than usual. As I have commented elsewhere, although having access to the world's media at the click of a mouse is a fascinating resource, it is also a mixed blessing. When I went to do the news pages for this issue I found so many things that I wanted to put in that I couldn't fit them into the normal space allocated in the magazines for news stories. So I made the decision to try and catch up with ourselves, which is why there are more news stories in this issue, and less reviews.

Newsfile Xtra

Sea Snake Stories

One of the things that I have noticed over the years that I have been chronicling Fortean zoological phenomena around the world is that such phenomena very seldom happen singly. Admittedly I am not the first person to have noticed this, nor would I wish to claim that this was so. No less a person than Charles Fort himself remarked upon it, and it could be argued that such waves of data are the basic building blocks of Fortean investigation.

Over the period of Samhain we have had a number of interesting stories about a group of animals that I have always, personally, found very interesting indeed.

Sea Snakes.

There are 62 known species, and one – *Pelamis platura* commonly known as the yellow-bellied sea snake may have the largest natural range of any reptile. Several species have been reported in enormous numbers. As a result of this it has been speculated that a sea snake species is the most numerous reptile in the world. This is almost certainly not true and this honour probably goes to the common lizard (*Zootoca vivipara*) which is found across Europe and Asia, even beyond the Arctic circle.

I have only seen living sea snakes once – at the state fair in Perth, Western Australia in 1968, although I did see a dead one in Hong Kong a few years earlier. They are remarkably difficult to keep in captivity. They appear to be intolerant of handling and also need a highly specialist diet. Species that have done relatively well in captivity include the ringed sea snake, *Hydrophis cyanocinctus*, which feeds on fish and eels in particular. *Pelamis platurus* has done especially well in captivity, accepting small fish, including goldfish. However, care

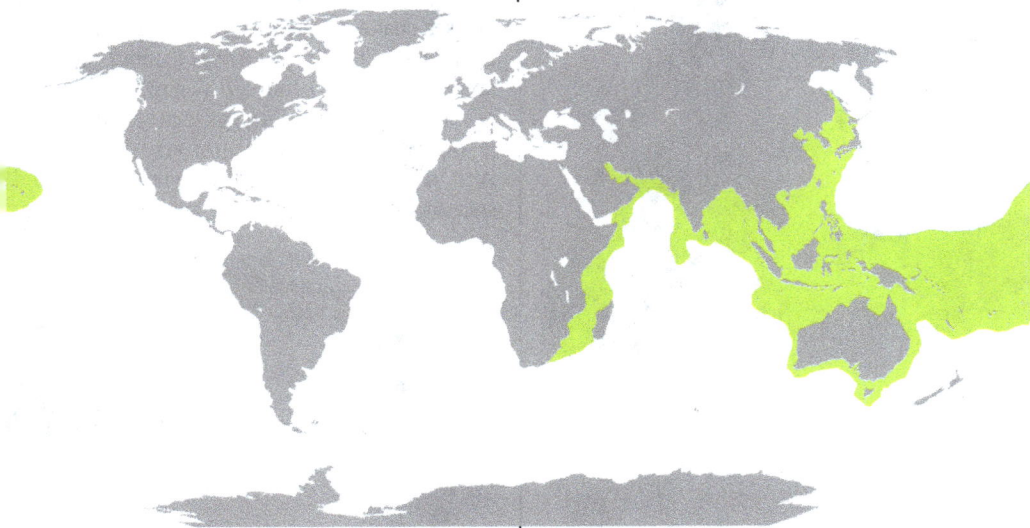

has to be taken to house them in round or oval tanks, or in rectangular tanks with corners that are well-rounded, to prevent the snakes from damaging their snouts by swimming into the sides.

There have been two recent accounts of out of

place sea snakes. The first is from the *Daily Mail* online of the 2nd of November which reports that: "A one and a half metre long Stokes' sea snake - which is known to live in

the tropical waters of Western Australia, Queensland and the Northern Territory - washed up in Manly Cove last week, much to the surprise of local resident Carole Douglas. The formidable marine snake, whose fangs are long enough to pierce a wetsuit, is highly venomous and with no known anti venom, the large ocean serpent is capable of delivering a painful and fatal bite".

The second report is, if anything, even more extraordinary, as you can see from the world wide distribution map we produced here, and comes from Ventura County, California and was reported by CNN on the 17th October. A surfer had captured footage of the sea snake lying on the beach. "It looked lethargic when I approached," Bob Forbes told CNN. "I touched it lightly and it started to move." Fearing that children might come across the aquatic snake, Forbes placed it inside a five-gallon bucket with some ocean water and alerted local wildlife experts. The discovery is a rare Southern California record, according to Greg Pauly, curator of herpetology of the Natural History Museum of Los Angeles County. "It was the northernmost sea snake ever documented in the Pacific Coast of North America," he said. The last sea snake species to be documented washed up on shore in 1972 in Orange County, which is about 100 miles south of Ventura County. "I never would have thought that a sea snake would wash up that far up north," Pauly said. And as we were going to press, a second specimen has just washed up in California.

Carl Marshall's Column

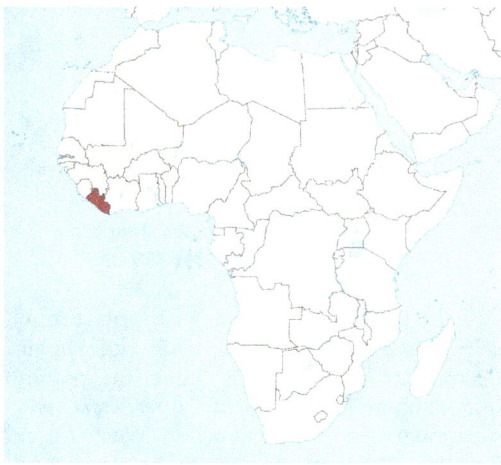

Dr Bernard Heuvelmans once postulated in his seminal work *On the Track of Unknown Animals* that the North West African country of Liberia could possibly conceal a formally undescribed species of pygmy rhinoceros cut off during the late Eocene Epoch (56 to 33.9 million years ago) and is now isolated in the countries mountainous coastal region.

The evidence for this unknown animal primarily came from native eye witness accounts reported to celebrated animal collector Major Hans Schomburgk when he visited Liberia in search of new species for zoo attractions. From these reports, one species, the pygmy hippopotamus, he eventually returned with successfully to Europe.

An Overview of the Effects of Liberia's Political Instability Within a Cryptozoological Research Framework

When Hans Schomburgk brought a pygmy hippopotamus back alive to Europe in 1913, and thus proved that at least one native legend was well-founded, he was probably more inclined to pay heed to others, one of which referred to a sort of pygmy rhinoceros which, according to the Kroo, also lived in the mountainous part of Liberia. At first he thought the tales must refer to the forest-hog, a big black pig with large tusks, a variety of which is in fact found in eastern Liberia. But he soon found that the Kroo made a clear distinction between the large forest-boar and the little mountain rhinoceros, and he eventually came to believe they were right. Unfortunately all his attempts to discover a specimen of this absolutely unknown species failed - it was a harder task than looking for a needle in a haystack, since here the needle was on the move, constantly eluding its pursuers... There is a good argument to show that this part of Africa may still conceal a pygmy rhinoceros.

Part six - The Terrors of Africa, On the Track of Unknown Animals (1958) p358-p359

While the common hippopotamus *Hippopotamus amphibius* was known to Europeans since ancient times, the pygmy hippopotamus *Choeropsis liberiensis* was unknown outside its range in West Africa until the 19th century. Due to their nocturnal forested existence, they were poorly known within their range as well. In Liberia, the animal was traditionally known as the "water cow". Early field reports of the species misidentified it as a wild hog! This mistake

might have parallels today with the alleged existence of the pygmy rhinoceros as we shall see later. The discovery of a new species of hippopotamus was so utterly unexpected in the 19th century that many scientists first tried to discredit the discovery by claiming the diminutive skulls being discovered to be either that of juvenile common hippopotamus, and then later when that theory was disproved, that they were an extinct animal by keeping museum specimens alongside prehistoric fossil species.

Considering Liberia's violent history, (see following section) coupled with the fact no recent scientist or explorer has bothered to investigate the claims, I believe this all poses the question whether a pygmy rhinoceros could still be living undiscovered in the country's remote and inaccessible mountainous region. The excitement generated since the publication by Dr. Bernard Heuvelmans highlighting the potential existence of this species has been largely superseded by civil unrest and to this day there have been no serious attempts made to obtain a type specimen of what might be West Africa's last unknown megafaunal mammal species.

The Atrocities of Liberia's Civil Wars - Systematic Slaughter, Rape, Child Soldiers and Cannibalism

The *Republic of Liberia* (also known as the Pepper Coast or Little America) is a country on the West African coast. Liberia literally means "Land of the Free". It is bordered by Sierra Leone to its west, Guinea to its north and the Ivory Coast to its east. Liberia covers an area of 111,369 square kilometres (43,000 sq mi) and is home to 4,396,873 people. English is the official language of Liberia

but only about 15% of Liberians speak it, the rest speaking other local languages of which there are about 20 indigenous dialects. Forests on the coast line are composed mostly of saline-tolerant mangrove trees while the more sparsely populated inland has forests opening onto a plateau of drier grasslands.

The climate of Liberia is equatorial, with heavy rainfall during the May-October wet season and harsh harmattan winds the remainder of the year. Liberia contains about forty percent of the remaining *Upper Guinean rainforest.*

Liberia, formally a colony of the *American Colonization Society* (ACS) declared its formal independence on Jul 26, 1847. The United States finally accepted and recognised Liberia's independence on February 5, 1862 and was the first African nation to gain its independence, although it

was not the only independent African state at the time. Liberia was established as a homeland for African Americans and ex-Caribbean slaves who came from the Caribbean islands and the United States with the help and support from the ACS. Liberia aided Britain for many years in its efforts to end the illegal West African slave trade. Early unrest in Liberia came when liberated slaves arrived from America and enslaved the original Liberian peoples using methods developed and learned from the west.

The modernisation of Liberia began in the 1940s following investment from the United States during World War II and economic liberalisation under President William Tubman. Liberia was an original founding member of *League of Nations*, *United Nations* and the *Organisation of African Unity*. In 1980 a military coup overthrew the *True Whig Party* leadership starting the beginning of a new wave of political instability.

Five years of military rule by the *People Redemption Council* and five years civilian rule by the *National Democratic Party* of Liberia were followed by two civil wars. These wars resulted in the deaths of between 250,000 and 520,000 people and devastated Liberia's economy for many years to come.

Political Unrest in the 20th Century

For a period of time in the early 20th century, Liberia became a U.S. protectorate. In the mid 20th century, Liberia gradually began to modernise with assistance from the USA. Both the *Freeport of Monrovia* and *Roberts International Airport* were built by U.S. personnel through the Lend-Lease programme during World War II. President William Tubman encouraged foreign investment in the country, resulting in the second-highest rate of economic growth in the world during the 1950s.

On April 12, 1980, a military coup led by Master Sergeant Samuel Doe of the *Krahn* ethnic group overthrew and murdered Present William R. Tolbert, Jr. Doe and the other plotters later executed a majority of Tolbert's cabinet and other Americo-Liberian government officials and *True Whig Party* members. The coup leaders then formed the *Peoples Redemption Council* (RPC) to govern the country. Samuel Doe, a strategic Cold War ally of the west, received significant financial backing from the United States during this time.

After Liberia adopted a new constitution in 1985, Samuel Doe was elected president in subsequent elections that were internationally condemned as fraudulent. On November 12, 1985, a failed counter-coup was launched by Thomas Quiwonkpa, whose soldiers briefly occupied the national radio station. Government repression intensified in response, as Doe's troops executed members of the *Gio* and *Mano* ethnic groups in Nimba County.

The *National Patriotic Front of Liberia*, a rebel group led by Charles Taylor, launched an insurrection in December 1989 against Doe's government with the backing of neighbouring countries such as *Burkina Faso* and *Cote d'Ivoire*, triggering the *First Liberian Civil War*. By September 1990, Doe's forces now controlled only a small area just outside the capital, and Doe was captured and executed that month by rebel forces. The rebels soon split into various factions fighting one another, and the *Economic Community Monitoring Group* under the *Economic Community of West African States* organised a military task force to intervene the spiralling crisis. From 1989 to 1996 one of Africa's bloodiest civil wars ensued, claiming the lives of more then 200,000 Liberians and displacing a million others into refugee camps in neighbouring countries. A peace deal between warring parties was reached in 1995 leading to Taylor's election as president in 1997.

Under Taylor's leadership, Liberia became internationally known as a pariah state due to the use of blood diamonds and illegal timber exports to fund the *Revolutionary United Front* in the *Sierra Leone Civil War*. The *Second Liberian Civil War* began in 1999 when *Liberians United for Reconciliation and Democracy*, a rebel group based in the northwest of the country, launched an armed insurrection against Charles Taylor.

The 2000's

In March 2003, a second rebel group, *Movement for Democracy in Liberia* began launching attacks against Taylor from the southeast of the country. Peace talks between the factions began in Accra in June of that year and Taylor was indicted by the *Special Court for Sierra Leone* for crimes against humanity that same month. By July 2003 the rebels had launched an assault on *Monrovia* - the capital city of Liberia. Under heavy pressure from the international community and the domestic *Women of Liberia Mass Action for Peace* movement, Taylor resigned in August 2003 and went into exile in Nigeria.

The Subsequent 2005 elections were internationally regarded as the most democratic in Liberian history. Ellen Johnson Sirleaf, a Harvard-trained economist and former *Minister of Finance*, was elected as the first female president in Africa. Upon her inauguration, Sirleaf requested the extradition of Taylor from Nigeria and immediately handed over to the SCSL to stand trial in the *Hague* for war crimes such as the illegal use of child soldiers. In 2006, the government established a *Truth and Reconciliation Commission* to address the causes and crimes of the civil war such as the unthinkable ritualistic cannibalism carried out by the rebels, often in public view and even recorded on film.

Some Unexpected Zoology and the Cryptozoology of Liberia revisited

The Little Mountain Rhinoceros
It wasn't until 1913, when big game hunter and animal collector Major Hans Schomburgk returned to Europe from Liberia with a live pygmy hippopotamus specimen that naturalists finally started to accept the possibility that relatively large undescribed animals could still be awaiting formal discovery in some of the remotest and inaccessible areas of sub-Saharan Africa. While searching for his illusive pygmy hippopotamus, Schomburgk was also informed of a strange mammal described by witnesses as a type of pygmy rhinoceros, which the native *Kroo* claimed was distinct from the large forest hog *Hylochoerus meinertzhageni* - a large black boar with prominent tusks well known to Liberian hunters. (see below)

Unfortunately, all Schomburgk's efforts to obtain a specimen of this mysterious and obviously extremely rare animal failed, and to this day, if indeed it does exist, no naturalist has ever seen, let alone collected and described, one of these so-called pygmy rhinoceroses. If the pygmy hippopotamus evolved in this part of Africa due to its prolonged isolation, could not a rare species of unknown rhinoceros have also evolved there in the almost inaccessible mountainous coast and now be evolved to it? Terrestrial animals which live on islands are generally smaller than the same or similar species living on the nearest continent, therefore the rhinos living on Java and Sumatra are smaller than those found in India and the Indochinese Peninsula. This phenomenon,

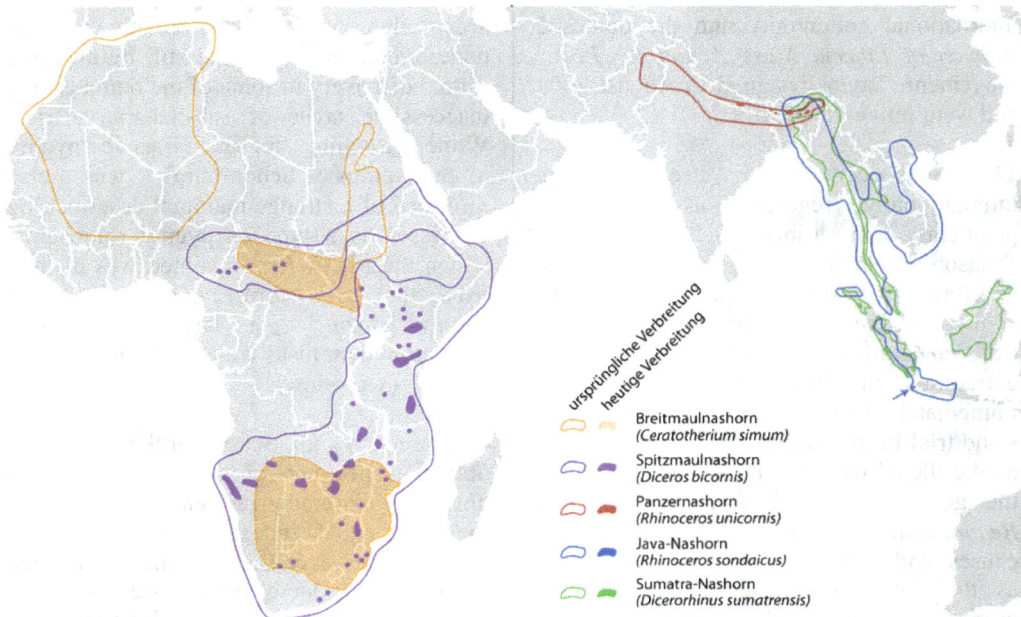

ursprüngliche Verbreitung
heutige Verbreitung

		Breitmaulnashorn *(Ceratotherium simum)*
		Spitzmaulnashorn *(Diceros bicornis)*
		Panzernashorn *(Rhinoceros unicornis)*
		Java-Nashorn *(Rhinoceros sondaicus)*
		Sumatra-Nashorn *(Dicerorhinus sumatrensis)*

when restricted territory influences the evolutionary process is called insular dwarfism; when a lack of viable sexual partners and dietary requirements both play a part and occurs not only on small islands but also cave systems, desert oases, isolated valleys and isolated mountain ranges. As a rule when mainland animals colonize islands small species tend to evolve larger bodies, and large species tend to evolve smaller ones.

To this day there remains a good argument that this part of Africa might still conceal small isolated populations of mountain rhinoceros similar in habits and possibly even appearance to the Sumatran rhinoceros *Dicerorhinus sumatrensis* (which it may or may not be related to). Coincidentally it is also the only known member of a genus, which being the least derived of all extant rhinos, shares more traits with prehistoric species such as

the woolly rhinoceros *Coelodonta antiquitatis* and is clearly the smallest known living species (approx 1,100 - 1,800 lb), no doubt due to its forest dwelling habits.

See conclusion for another theory as to a possible identity of this mysterious animal.

The Gbahali

The *Gbahali* is not thought of as a mythical creature but a living animal among Liberian natives that they have hunted and even consumed. It is considered to be an immense reptile similar in appearance to a crocodile but with a few subtle differences such as a shorter snout and proportionally larger limbs. The *Gbahali* are thought to have a nose to tail length of up to thirty feet making them only moderately longer than the largest known species - the salt water crocodile *Crocodylus porosus*. When shown cards displaying images of both extant and extinct species the hunters compared the

White rhinoceros (*Ceratotherium simum*)

Black rhinoceros (*Diceros bicornis*)

Rhinocerotidae

Sumatran rhinoceros (*Dicerorhinus sumatrensis*)

Javan rhinoceros
(*Rhinoceros sondaicus*)

Indian Rhinoceros
(*Rhinoceros unicornis*)

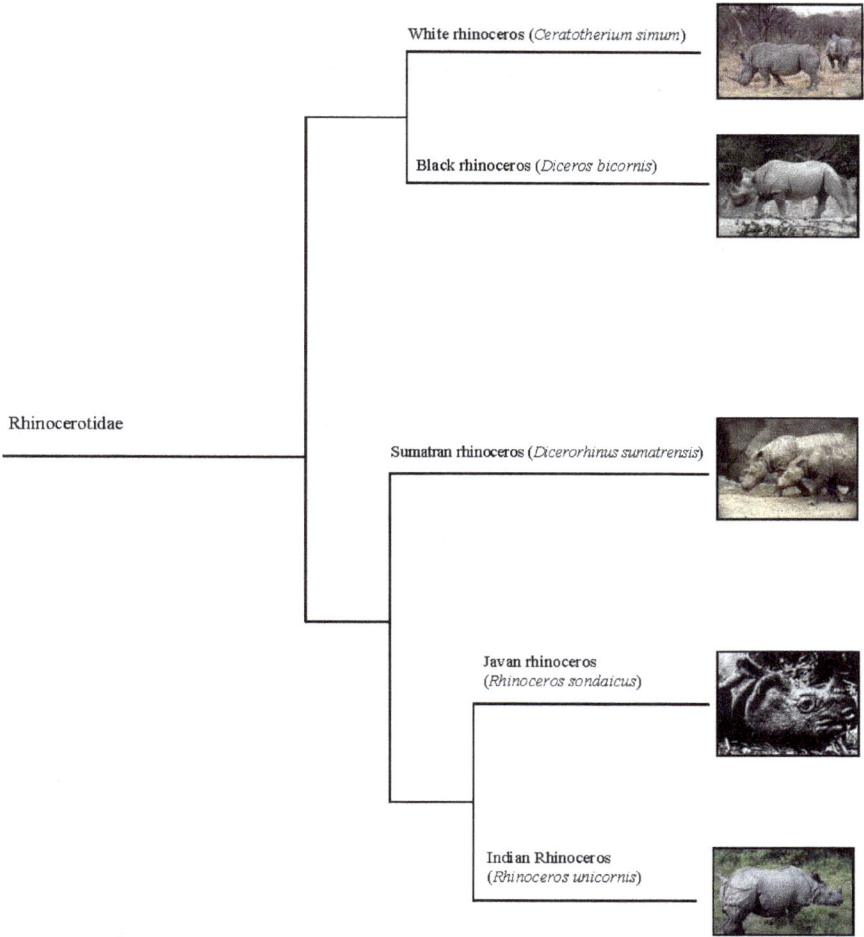

CLADOGRAM OF KNOWN LIVING
RHINOCEROS SPECIES

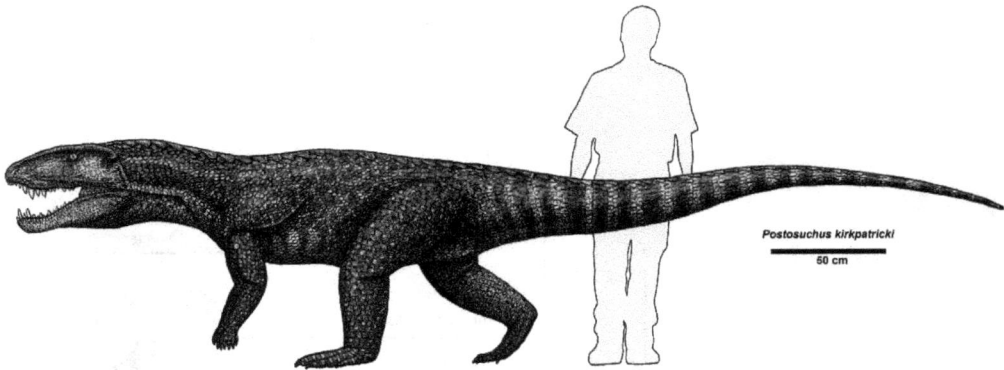

Postosuchus kirkpatricki
50 cm

likeness of the *Gbahali* to the now extinct *Postosuchus*. (above)

Strange Sloth-Like Animal

There are numerous stories about a strange "sloth-like creature" which is supposed to be found in Liberia, however informers have been unable to sufficiently describe this strange unknown animal in any detail to form a clear picture of it.

The Temu

The *Temu* are thought to be small bipedal primates covered with long hair especially on the head and nape. They display many hominid features but are much smaller than the average Liberians. Whistling sounds seem to be their primary method of communication. They are widely thought to dwell in the dense forests around Robertsport in Western Liberia.

Unidentified Caterpillar Plague

A devastating caterpillar plague that is ravaging parts of northern Liberia are not armyworms *Spodoptera exempta* as previously believed, but an unidentified species. A UN emergency coordinator told the BBC that the strange insects in Liberia and Guinea were very different from known species of armyworms. The coordinator said experts had noted the insects have distinct feeding patterns, life cycle, habits, movement and appearance. Entomologists are studying the pest to find a way of controlling the swarm which has so far affected 400,000 residents. As well as devouring crops, the infestation has polluted water sources with faeces. *UN Food and Agriculture Organisation* emergency coordinator Tim Vaessen said entomologists had realised the species were not armyworms during a field trip to *Bong County* in Liberia in late January 2009. "While armyworms feed on ground cereal like millet, rice or sorghum", Mr Vaessen said, "the unidentified caterpillars favour munching the leaves of the Dohama tree".

Mr Vaessen told the BBC News website: "Our experts in the field spoke to villagers who said they had seen this type of caterpillar before. They said they had put leaves under trees and

burned them to suffocate the caterpillar with smoke". However the villagers also said they had never seen these insects in such large numbers before.

The fact these voracious insects were ethno-known to locals and unknown to the scientific community must surely make them true microcrypids!

Environmental
issues and Problems Visiting Liberia Today

Since the civil wars the flora and fauna of Liberia has been steadily reviving. During previous conflicts the bush meat trade flourished threatening many rare species such as the pygmy hippopotamus. Many organisms were decimated although on the whole extinction has been avoided. The pygmy rhino however (if indeed it is a true rhinoceros) appears to have escaped the slaughter as there is no evidence that it was hunted for bush meat and suffered the same fate. Although its horn could have potentially created a market for the indigenous peoples spiritual superstitions and for medicinal purposes (alleged properties of rhino horn), this seems not to have occurred, perhaps because the diminutive animal may have diminutive horns! When interviewed by Schomburgk Liberian hunters were unclear whether the animal has two horns like both the African species and the Sumatran species (which seems more probable) or a single horn like the Indian and Javan animals. Or for that matter whether the appendage is pronounced or not!

Liberia has a history of violent civil wars. At present it is less dangerous but has again become the world's focus due to the recent outbreak of the Ebola virus which has claimed the lives of thousands of people and threatened a global pandemic. This has resulted in more discouragement of free movement and, once again, the prevention of zoologists and environmental scientists to monitor species effectively through field research hampering progress in this area. The rehabilitation of Liberia and its population is ongoing but is unfortunately still tainted by its violent history; such as the many child soldiers whose rehabilitation was introduced but proved unsuccessful. These individuals are now adults with some having a continued predisposition towards sporadic violent acts.

Conclusion.
Liberia's past, and the decimation that the civil wars caused, has prevented scientists to implement serious prolonged research into this area and bring the mystery of the pygmy rhinoceros to an adequate conclusion. This must be addressed in future! It seems to me the theory of an unknown species of pygmy or dwarf rhinoceros surviving in Liberia's mountainous coast is by no means impossible considering the inaccessible habitat this animal is claimed to live in. However, I also see another maybe more plausible possibility.

Considering Liberian hunters who claim to have killed and eaten this animal can't describe its horn in any detail this must be taken negative evidence. Either it hasn't much of in the way of horns (or horn) or its not a real rhinoceros at all! Maybe Schomburgk was right the first time and this mystery African animal is actually a large unknown species of **Suidae** related to, but genetically and morphologically distinct from, the known black forest hog *H. meinertzhageni.* Personally

this latter theory seems the more plausible and if validated would still ultimately prove a great success for cryptozoology!

Special thanks
To Richard Muirhead for his tireless research, Frances Marshall for proof reading, and to Maureen Ashfield for her wonderful illustration of a Liberian pygmy rhinoceros.

Sources and suggested further reading.

* **Delmont, J.** *Catching Wild Beasts Alive.* (London: Hutchinson, 1931).
* **Freeman, R.** ORANG-PENDEK: *Sumatra's Forgotten Ape.* CFZ Press (2011).
* **Heuvelmans, B.** *Les betes humaines d' Afrique.* (Paris: Plon, 1980).
* **Heuvelmans, B.** *On The Track Of Unknown Animals.* Rupert Hart-Davis (1958) 2nd ED.
* **Kingdon, J.** *The Kingdon Guide To African Mammals.* Academic Press Limited, London (1997).
* **Ley, W.** *The Lungfish And The Unicorn.* (New York: Viking 1941).
* **Orthneil Forte, D.** *Famous Liberian Folklore: A Collection of Short Stories Across Liberia.* (Createspace, 2013).
* **Shuker, K.** *Extraordinary Animals Worldwide.* (London: Robert Hale, 1991).
* **Wikipedia the free encyclopedia.** en.wikipedia.org

Carl Marshall works at Stratford Butterfly Farm and is a fine field naturalist. Over the past couple of years he has become a very enthusiastic member of the CFZ, and his quasi-Fortean view of British natural history fits in perfectly with my own. He was, therefore, the perfect choice as a columnist for the brave new *Animals & Men*, and we are proud to have him aboard.

CORINNA DOWNES

The most disturbing news in the months since the last published issue of *Animals & Men*, was the publication in October of the latest annual revisions of birds on the IUCN Red List. In it, the number of UK bird species considered to be facing the risk of extinction has doubled to eight, with a further 14 considered to be near threatened (this meaning that any further downtrend in their status may well see them added to the red list also).

As the RSPB reported: "Martin Harper is the RSPB's Conservation Director. He said: "Today's announcement means that the global wave of extinction is now lapping at our shores. The number of species facing extinction has always been highest in the tropics, particularly on small islands. But now the crisis is beginning to exact an increasingly heavy toll on temperate regions too, such as Europe.

"The erosion of the UK's wildlife is staggering and this is reinforced when you talk about puffin and turtle dove now facing the same level of extinction threat as African Elephant and Lion, and being more endangered than the Humpback Whale."

"The global revision also captures the crisis facing other birds around the world, including vultures where several African species have been listed as Critically Endangered – one step away from facing global extinction. In Africa, vultures are facing persecution and they are regularly poisoned or trapped.

"Examining the list of changes among the UK's birds to this year's red list, several themes emerge, including: deterioration in the fortunes of some seabirds, such as puffin and razorbill; an ongoing and increasingly intense threat to wading birds, such as godwits, Curlew, Oystercatcher, Knot and Lapwing; and an increasing deterioration in the status of marine ducks, such as Common Eider, joining Velvet Scoter and Long-tailed Duck as species of concern.

- "**Turtle Dove:** formerly the Turtle Dove was a familiar summer visitor to much of Europe including south-east England. Declines across Europe exceeding 30 per cent over the past 16 years have seen its threat status rise from Least Concern to Vulnerable. Scientists from the RSPB and other BirdLife International partners are trying to establish the reasons for the decline in the UK and Europe. The decline in the UK has been particularly high, with more than nine out of every ten birds being lost since the 1970s.

- "**Atlantic puffin:** Globally, this seabird is concentrated in Europe. Although its global population remains in the millions, breeding failures at some key colonies over recent years have been worryingly high, with many fewer young birds being recruited into the breeding population. These declines mean the species has been assessed as Vulnerable. Large declines have been reported in Iceland, the Faroe Islands and Norway, which together hold 80 per cent of the European population. In the UK, there have been significant losses on Fair Isle and the Shetland Islands, but elsewhere in the UK this seabird seems to be doing well.

- **"Slavonian grebe:** The Slavonian Grebe occurs across North America, northern Europe and northern Asia. The bulk of this waterbird's population occurs in North America where it has undergone a large and significant decrease. This decline has triggered the inclusion of the Slavonian Grebe on the list of species evaluated as Vulnerable. However, new information collated from across Europe suggests the Slavonian Grebe is declining here too. In the UK, the number of nesting Slavonian Grebes, all in the Scottish Highlands, have declined although those wintering round the UK's coasts have increased.

- **"Pochard:** The Pochard occurs widely across Europe and Asia. Recent information collated from across Europe indicates that this duck has declined significantly in recent years and that this decline is ongoing. Globally, the Pochard has been uplisted to Vulnerable. In the UK, the numbers of nesting Pochard and the number of wintering individuals have declined markedly.

- **"UK and European wading birds:** The addition of Knot, Curlew Sandpiper, Bar-tailed Godwit, Oystercatcher, and Lapwing to the list of Near Threatened Species is troubling for this group of birds, especially as the newly-listed species join others, such as the Curlew and Black-tailed Godwit, which have been listed as Near Threatened in previous assessments. The Lapwing is a species which nests widely across Europe, including formerly most parts of the UK. Information from across Europe indicates that this bird has declined significantly in recent years and that this decline is ongoing. The Bar-tailed Godwit, Knot and Curlew Sandpiper are species which nest in the high Arctic and spend the winter on coasts further south, including in Britain. Declines in parts of their wide range, such as Siberia or eastern Asia, has led to the listing of these species as Near Threatened.

- **"African vultures:** six of Africa's 11 vulture species have had their global threat status upgraded to a higher level. Four of the species have been uplisted to Critically Endangered the highest category of threat before extinction. The main causes of the drop in African vulture populations are thought to be indiscriminate poisonings, where the birds are drawn to poisoned baits meant for predators such as lions or hyenas. The use of vulture body parts in traditional medicine and deliberate targeting by poachers trying to prevent the authorities being alerted to illegal ivory poaching or other wildlife crimes are also significant reasons for the illegal killing of vultures.

- **"UK and European seaducks:** Although the Eider has a large global range across North America, northern Europe and northern Asia, new information from Europe suggests that this marine duck has declined significantly in recent years and that this decline is ongoing. The Eider is one of a number of coastal ducks which have declined in recent years. In the UK, the Long-tailed Duck and the Velvet Scoter have both been listed as Vulnerable and Endangered respectively. Although this year's revision of the red list has seen the Velvet Scoter's status dropped to Vulnerable.

RSPB - 29 October 2015"

http://www.rarebirdalert.co.uk/v2/Content/RSPB_Puffin_and_Turtle_Dove_join_the_red_list.aspx?s_id=170171181

Laysan Albatross
It was reported at the end of November that a Laysan albatross, estimated as being sixty-four years old, returned to Midway Atoll National Wildlife Refuge – the world's largest nesting albatross colony located north-west of Hawaii. Wisdom, as she is known, was spotted with a mate and it is expected that another chick will be raised, bringing her total number of chicks to around 36. Wisdom is the oldest living tracked bird, having been first tagged with a band in 1956, but Laysan albatross do not return to breed until they are at least five years old, so her age is only an estimate; she could be older.

Laysan albatrosses typically lay one egg a year, spending more than 130 days incubating it. The birds, which can have a wingspan of up to 7ft (2 meters), can forage hundreds of miles out to sea for food such as squid. It's thought Wisdom has notched up around 6m ocean miles of flight.

Dan Clark, manager of the wildlife refuge, said Wisdom had become a "symbol of hope and inspiration" in the face of a precipitous decline in seabirds. An estimated 70% of the creatures have disappeared worldwide since the 1950s.

"We are a part of the fate of Wisdom," Clark said, "and it is gratifying to see her return because of the decades of hard work conducted to manage and protect albatross nesting habitat."

Bret Wolfe, Clark's deputy, added: "It is very humbling to think that she has been visiting Midway for at least 64 years. Navy sailors and their families likely walked by her not knowing she could possibly be rearing a chick over 50 years later. She represents a connection to Midway's past as well as embodying our hope for the future."

Source: The *Guardian*

http://www.theguardian.com/environment/2015/nov/30/worlds-oldest-tracked-bird-wisdom-laysan-albatross-midway-atoll

Acadian Flycatcher
On 22nd September, an American flycatcher species was spotted at Dungeness in Kent. A slight bird, they weigh less than half an

ounce, it was definitely out of its usual habitat; they would be in a forest in eastern North America for breeding or in a forest in northern South America to winter, or in transit between them. It should not have flown across the Atlantic to end up on a shingle beach in Kent. After photographs were taken and investigated, it was pronounced an Acadian flycatcher; a previous specimen being found dead in Iceland in 1967 with no others having been seen this side of the Atlantic. This was a first for Britain.

Experts believe the Acadian flycatcher got caught up in a fast-moving Atlantic weather system that took it from the eastern US to Dungeness in Kent.

Dungeness birdwatcher Paul Trodd, who took photographs of the bird, said: "We have managed to pick up some of its poo for DNA analysis, which will go towards confirming the bird's identification. "When the sun came out this afternoon and the bird was just moving around close to our feet was simply amazing."

BOURC chairman Professor Martin Collinson, a senior lecturer at the University of Aberdeen, described the DNA sampling technique and the results.

"When the fecal sample arrived at University of Aberdeen we digested it using enzymes to make a soup of the bird's DNA and then used a polymerase chain reaction (PCR) reaction to make enough of the mitochondrial DNA for sequencing. The problem with poo samples is that the DNA is rapidly degraded into tiny fragments, so the normal PCRs we use to isolate big pieces of DNA from feathers often do not work, and true to form they did not work for the Dungeness bird. We therefore designed a PCR protocol that would allow us to sequence small fragments of the DNA mitochondrial cytb gene from any Empidonax species."

"The sequence we got was fed into the online database of every bird that has ever been sequenced and was identical to 3 of the 4 individuals of Acadian Flycatcher from USA and Mexico that were in there, and 1 base pair different from the fourth bird,. In contrast all other species of Empidonax, including Alder and Willow, were at least 8-9% different. On this basis, we can say with 100% confidence that the Dungeness bird was genetically an Acadian Flycatcher."

Sources: The *Guardian*
http://www.theguardian.com/
environment/2015/sep/26/birdwatching-
mega-month-september-acadian-flycatcher
The *Telegraph*
http://www.telegraph.co.uk/news/earth/
wildlife/11885236/Rare-North-American-
bird-arrives-on-British-shores-for-the-first-
time.html
Rare Bird Alert

http://www.rarebirdalert.co.uk/v2/Content/Finders_in_the_Field_Acadian_Flycatcher_Dungeness_Kent_September_2015.aspx?s_id=383972835

Thick-billed Warbler

A Thick-billed Warbler (*Iduna aedon*) was discovered at Quendale on Thursday 24th September, and if confirmed would be the sixth for both Britain and Shetland.

This warbler is an Old World bird that breeds in temperate east Asia. It is migratory, wintering in tropical south east Asia. It is a very rare vagrant to western Europe.

Source: BirdGuides

Blyth's pipit

A Blyth's pipit (*Anthus godlewskii*) was found on the saltmarsh at Stiffkey in Norfolk in September. This represents the third-ever record for Norfolk and the first new Blyth's found in the UK this year.

This is a medium-sized passerine bird which breeds in Mongolia and neighbouring areas. It is a long distance migrant moving to open lowlands in southern Asia, and is a very rare vagrant to western Europe.

This bird was named after the English zoologist Edward Blyth.

Source: BirdGuides

INTERESTING NEW SPECIES

New species find in Central Otago confirms link between Australian and South American shorebirds. We all know that birds evolved from dinosaurs, but it is uncertain as to what occurred next in the grand scheme of things. Today, shorebirds, or waders as they are also known live in a wide variety of environments worldwide, from the Himalayas to Antarctica, and have long been a subject of evolutionary discussion, but due to a poor fossil record, these questions remain largely unanswered. But a new article published in *Journal of Systematic Palaeontology* reports on a new piece in this evolutionary puzzle

An international team of New Zealand and Australian-based scientists, including researchers at Canterbury Museum, have confirmed that a 19-16 million-year-old shorebird fossil, discovered in Central Otago, New Zealand, belongs to a group of small birds including the Australian plains-wanderer and the South American seedsnipes.

Named after a 'mystery bird' in Māori mythology and in honour of New Zealand-based ornithologist and ecologist David Melville, this new species of small wading bird has been christened *Hakawai melvillei*. It lived about 19 million years ago during the Miocene epoch, around an ancient subtropical lake on the edge of a floodplain. Although *Hakawai melvillei* is extinct, its ancient lineage and close relationships show how all these birds have a common ancestry in East Gondwana, before the landmass subsequently split up and New Zealand became isolated and presents evidence that these birds were ancestral waders.

Source: http://www.sciencedaily.com/releases/2015/10/151020091834.htm

'Ghost' kingfisher rediscovered in Pacific - then killed as specimen

It was reported in September that a male Guadalcanal moustached kingfisher had been photographed for the first-time ever on the Solomon Islands – the species had not been seen since the 1950s.

Scientists in the remote highlands of Guadalcanal in the Solomon Islands, have been on the island surveying the endemic biodiversity and working with local partners to create a protected area.

Director of the project involving scientists surveying biodiversity on the island, Chris Filardi, wrote on his blog: "After several days of work, it is clear we are on the shores of an island in the sky. Species we encounter here are of two worlds – one that descends to the humid, coastal plain and another that rises into the cool, cloud-raked mountains of Tetena-Haiaja. Just as the white sands of an island beach divide land and sea, the ascending Chupukama ridge marks the transition from a world of known lowland organisms to a sky island filled with scientific mystery.

"In the western Pacific, first among these 'ghost species' is Moustached Kingfisher, a bird I have sought for nearly 20 years. Described by a single female specimen in the 1920s, two more females brought to collectors by local hunters in the early 1950s, and only glimpsed in the wild once. Scientists have never observed a male. Its voice and habits are poorly known. Given its history of eluding detection, realistic hopes of finding the bird were slim."

The Uluna-Sutahuri people who live in the forests call the bird 'Mbarikuku'. BirdLife International lists this species as "Endangered on the basis of a very small estimated population which is suspected to be declining, at least in part of its range. However, further research may reveal it to be more common."

First photographs of the species, and the first recordings of its call, have now been obtained, and it is hoped that the green-backed females will also be photographed. Not only that, but the bird and its habitat are reportedly still thriving on the island. The male was taken as a specimen (killed) for further study at the American Museum of Natural History, an action that appears to be standard in such circumstances among American museums and research institutions. However, this act is ever-more controversial in the era of digital photography and DNA sampling, and caused an uproar from many quarters.

Source: Birdwatch News Archive
http://www.birdwatch.co.uk/channel/
newsitem.asp?c=11&cate=_16088

The Wilson's warbler (*Cardellina pusilla*) has not been seen in Britain for 30 years but during October, one was spotted at Port Nis, Lewis. This warbler breeds across much of northern North America, and although its range extends eastwards in eastern Canada, it is more common further west. Its westerly migration route seems to make it an unlikely vagrant in Europe. Ireland was lucky to have had a visit from one in 2013.

At first thought to be a yellowhammer, upon closer investigation a bird that appeared on Papa Westray, Orkney, was confirmed as a chestnut bunting (*Emberiza rutila*). This first for Britain bunting is fairly small and found in eastern Asia,

breeding in Siberia, northern Mongolia and north-eastern China. It migrates long-distance and winters in southern China, south-east Asia and north-east India. Although there have been a number of records from Europe, some of them are considered to be escapes from captivity rather than genuine vagrants. There are nine accepted records in the Western Palearctic: in the Netherlands in 1937, Norway 1974 and 2010, Malta 1983, Slovenia 1987, Finland 2002, France 2009 and 2014, and Hungary 2011. All of these records fell between 24th September and 12th November, with six in October. Chestnut Bunting is currently only on Category E of the British List, with eight rejected records of presumed escapes; on Foula (Shetland Isles) on 9th-13th June 1974, Isle of May (Fife) 11th June 1985, Fair Isle (Shetland) 15th-16th June 1986, Bardsey Island (Gwynedd) 18th-19th June 1986, Out Skerries (Shetland Isles) 2nd-5th September 1994, Salthouse (Norfolk) 30th May-1st June 1998, Whitburn (County Durham) 17th-20th May 2000, and Fair Isle (Shetland) 4th-7th September 2002.

Source: Rare Bird Alert

There was a crooked bird that lived on a crooked spire
Chesterfield, the home of the Church of St Mary and All Saints with its crooked spire, welcomed a Eurasian crag martin on 8th November. A committee member for the Sheffield Bird Study Group said, "The birding community is huge and the excitement of finding or seeing a rare species of bird is what drives many people to travel the country, and in fact the world, in order to increase their species count."

The Eurasian crag martin, normally found in

Southern Europe, northwest Africa and Southern Asia, is extremely rare in the UK. "This is the first bird for Derbyshire, first for the Sheffield recording area and I believe only the tenth bird seen in the UK. Hence it is labelled "MEGA" by the birding fraternity," he added.

Sources: http:// www.derbyshiretimes.co.uk/news/ grassroots/birdwatch-tracking-the-crag-martin-in-chesterfield-1-7577107#ixzz3ub88NCEa

Source: Rare Bird Alert

Green-winged macaws back in Argentina

Macaws have long been persecuted by humans due to their colourful plumage, but now green-winged macaws have been released in northeastern Argentina, after

being absent for almost two hundred years. There were at least two species: the glaucous macaw (*Anodorhynchus glaucus*), which became globally extinct, and the green-winged macaw *(Ara chloropterus)*, which also disappeared from the region, where they inhabited fields with jungle-islands between estuaries, and palm and gallery forests along the waterways.

Due to the perilous nature of the green-winged macaw's survival in Corrientes, a recovery project was begun which focusses on using captive green-winged macaws originating from several zoos and breeding centres around the country. These birds form the "Ecological Complex of Aguará," located in the province of Corrientes, where groups of individuals are consolidated and all health checks are performed to rule out diseases that may be spread in the wild following the release. Before their release, the birds spent several weeks in an acclimation aviary in the Cambyretá area, proving the northern access to the Esteros del Ibera.

The Conservation Land Trust is financing most of the project thanks to the donation of a European philanthropist, and bringing its previous experience in wildlife reintroduction projects in Ibera. The CONICET scientists contribute their knowledge on the ecology of these birds and their reintroduction. The Directorate of Natural Resources of Corrientes provides the Center Aguará facilities, where the macaws are kept before being transferred to Ibera. Also, several groups of volunteers, including scouts, schools and club birders help to disseminate information on the project, and contribute their observations of animals in the field. Through this initiative, Argentina regains its first extinct species from the ex-situ management of wild bird specimens, and will continue working on their recovery through intensive management.

More information:
www.proyectoibera.org/guacamayo

Source: Birdlife International

http://www.rarebirdalert.co.uk/v2/Content/ Birdlife_Green_winged_Macaw_back_in_

Argentina.aspx?s_id=170171181

Makira Moorhen
The Makira moorhen was last recorded in 1953, but due to sightings being reported in more recent times, it is thought unlikely to have gone extinct, although any remnant population is likely to be tiny. Solomon Islands Community Conservation Partnership (a potential in-country partner for BirdLife) in conjunction with Mark O'Brien at BirdLife with support from the Critical Ecosystem Partnership Fund have co-ordinated a new search for the bird.

An interview with a local hunter provided plausible evidence of, and detailed description of, a bird that sounds like a Makira moorhen being sighted, and killed, in East Makira less than 3 years previous. A second, rather more sketchy and less detailed story of a Makira moorhen being sighted in 2011 was also relayed ot the team. These add to 3 other reports from hunters of birds in similarly wild, remote areas of the East Makira forest this century.

Camps were established at three separate locations during the fieldwork – the third of which was located in good quality native forest. Unfortunately, the weather deteriorated during the time at the third camp, making observations difficult. However, automatic cameras were set – and the researchers are looking forward to checking any resulting images.

The team located a nest and single egg of a yellow-legged pigeon, which is a declining and globally-threatened species, and is considered to be rare wherever it occurs, although it is restricted to the Solomon Islands and Bismarck Archipelago, apart for Makira where it is classed as locally uncommon. Volunteer, Reuben Tako, located the nest, and the continued presence of the pigeon in the area might indicate that ground predators are not at a high density within the Makira forest. This is an inference that bodes well for the continued existence of a flightless ground-dwelling species, such as Makira moorhen.

Birdlife International
09 November 2015
http://www.rarebirdalert.co.uk/v2/Content/Birdlife_Search_for_the_lost_Makira_Moorhen.aspx?s_id=170171181

Mass starling drownings puzzle scientists
It was published recently, in the journal *Scientific Reports*, that the mysterious cause of death amongst groups of young common starlings (*Sturnus vulgaris*) was by drowning, which is rare amongst wild birds, and even then usually only involves one rather than multiple birds, according to a team of scientists led by international conservation charity the Zoological Society of London (ZSL). However, starlings have been seen to drown in groups of around 10, which has led to scientists investigating these unusual incidences.

Dr Becki Lawson, lead author and wildlife veterinarian at ZSL, commented: "Drowning appears to be a more common cause of death amongst younger birds, as they may be inexperienced in identifying water hazards. This combined with the fact that Starlings are a highly social species could potentially explain why multiple birds drown together."

"Members of the public have been instrumental in bringing this unexpected cause of Starling mortality to our attention by reporting these incidents. With Starling numbers declining in general across the UK, we need to learn more about how and where these phenomena happen, in order to better understand why," Dr Lawson explained.

Rob Robinson, co-author and Associate Director of Research at the British Trust for Ornithology (BTO) said: "Starlings are a Red-listed species in the UK, under threat from issues including loss of nesting sites and a lack of insect food sources – so much so that their population has declined 79 per cent in the past 25 years. Whilst drowning is an unexpected cause of death, it's not thought to be a conservation threat as – fortunately – these incidents are currently relatively rare. However, we still need to better understand factors such as disease that might be contributing to this decline. We would therefore ask people to keep up the good work by reporting incidents of Starling death, whatever the apparent cause, via the **Garden Wildlife Health website**."

Water can be a valuable resource for wild birds, particularly during the summer months. Providing water sources such as ponds or bird baths is still recommended as a way to support garden wildlife. However, experts also recommend adding a sloping exit or ramp to water features, in order to help birds and other animals easily access and exit water sources.

If you spot sick or dead wildlife in gardens you can help scientists learn more about their cause by reporting these incidents via the **project website**.

02 December 2015
http://www.rarebirdalert.co.uk/v2/Content/Mass_starling_drownings_puzzle_scientists.aspx?s_id=170171181

World's biggest seabird tracking database shows their incredible journeys
One of the biggest marine conservation collaborations in the world, the Global Seabird Tracking Database, has just passed 5 million data points. Established in 2003, the database - originally called 'Tracking Ocean Wanderers' - brought together, for the first time, data on the movements of 16 species of albatross and petrel. From albatrosses to penguins, the database now holds more than five times as many species, provided by over 120 research institutes.

Some insights include:

The individual bird tracked for the longest period of time was a juvenile Tristan albatross, from 21 Dec 2013 to 07 Jan 2015, during which time it travelled 186,684 km. That means travelling nearly 500km every day for 383 days.

More than half of the data relate to threatened (i.e. Critically Endangered, Endangered or Vulnerable) or Near Threatened seabirds - species for which conservation efforts are most pressing.

Dr Cleo Small, who has led BirdLife's work to get seabird bycatch conservation measures adopted by the high seas tuna management organisations, said: "It's hard to imagine that we could have convinced fisheries managers to put seabird bycatch on the agenda without the seabird tracking data. It's truly been the cornerstone of our efforts to prevent the tragedy of seabird bycatch".

Professor John Croxall, who helped establish the database in 2003, added: "Our first meeting brought together nearly all the seabird tracking data that existed at that time, covering 16 species of albatross. It's fantastic to see the database grow so dramatically, thanks to the willingness of the seabird research community to share their data in the name of furthering marine conservation".

All the tracking data can be viewed at www.seabirdtracking.org

http://www.rarebirdalert.co.uk/v2/Content/Birdlife_Worlds_biggest_seabird_tracking_database_shows_their_incredible_journeys.aspx?s_id=170171181

Western rufous turtle dove

A Western rufous turtle dove (*Streptopelia orientalis*) was discovered in a garden at Scalloway, Shetland on 25th November. There are three previous records confirmed to have been of the form *meena*, with the Orkney individual of 2002 being the only one easily observed. There are three previous accepted British records of *meena*:

- **2003:** first-winter, Hill of Rattar, Caithness, 5 December to at least 24 March 2004

- **2002:** juvenile to first-winter, Stromness, Orkney, 20 November to 20 December

- **1975:** juvenile to first-winter, Spurn, E Yorks, 8 November

-

The last twitchable Oriental Turtle Dove in Britain was a well-watched first-winter of the subspecies *orientalis*, present in Chipping Norton, Oxon, from 15 December 2010 to 9 May 2011.

Both the names Oriental turtle dove and rufous turtle dove have been used for this species. It has been suggested that the name rufous turtle dove should be used for the western form *meena*, and Oriental turtle dove for the nominate *orientalis*.

In 1994, the British Birds Rarities Committee reviewed the British records of this species and concluded that three of the eight accepted records should no longer stand. The Oriental turtle dove has occurred a number of times since, including a well-watched bird in Oxfordshire in February 2011, which was found in a birdwatcher's garden. The house owner used the opportunity to charge a £5 fee to view the bird from his kitchen window and up to 500 people queued to see the bird.

American Bittern

In County Cork an American bittern (*Botaurus lentiginosus*) was found at the lake below Castlefreke House, Owenahincha. The last Irish record was in 1990, when one was killed by a dog in County Wexford, and you have to go back to 1875 for all three of the previous County Cork records, one of which was also found (and shot) on this date. breeding in Canada and the northern and central parts of the United States, and wintering in the southern United States, the Caribbean islands and parts of Central America.

Source: Rare Bird Alert

For those of you not aware, as well as this column in *Animals & Men,* Corinna writes a daily Fortean bird blog which can be found as part of the CFZ Blog Network, but also as a stand alone site at:

http://cfzwatcheroftheskies.blogspot.com/

2015 Butterfly Report

In Issue 53 we presented a round up of news about butterfly vagrants in the UK. Now we bring the story up to date

The 2015 butterfly season is now well and truly over. In fact, if you want to be nit picking about it, it isn't, because there will be occasional sightings of butterflies that have awoken early from hibernation throughout the winter months, but you know what I mean.

This was a far less spectacular year than previous years have been, but still there were items of interest, most notably the monarchs that turned up from May - October. Sadly, this is one of the species that people in the wedding industry release at weddings, so one cannot say for sure that any of them are genuine vagrants.

Surely this is illegal under the Wildlife and Countryside Act? Even if it isn't, the symbolism of paying hundreds of pounds to release non native insects into a country where they cannot breed and will die slowly and unpleasantly, would seem to be highly inappropriate for what is meant to be a joyous occasion.

I think that it is ostentatious, cruel and vulgar, and would like to see it banned.

But what do I know? I am just a miserable old git with a bad attitude.

Lampides boeticus is one of the most common and wide spread butterflies in the world, although it is a very rare visitor to the United Kingdom. The UK and Western Europe are at the Northern most limit of range, and so, historically it has been a vagrant. Unlike many native species of butterfly it is continually brooded which means that generation lives out its entire lifecycle in a short time, repeated throughout the year and – unlike all British species of butterfly it doesn't hibernate during any of its stages. Unless it changes its lifestyle dramatically which is not impossible, although it is highly unlikely, it will not be able to survive a British winter.

There is, however, a interesting precedent which provides a scenario which could – if extrapolated – could suggest a way in which this could happen.

In the 1970's two populations of guppy (*Poecilia reticulate)* a popular aquarium fish, whose natural range is in North East South America, established two colonies in the UK despite being undeniably a tropical species

According to Sir Christopher Lever, the world's expert on the subject of naturalised animals, during the 1970's there were at least two colonies of these hardy little tropical fish living wild in the United Kingdom. But how is this? you might well ask. The more eco-concerned of you might start to rant about the abominable effects of global warming, but in fact the tale has a far less portentous explanation. Between July 1966 and February 1968, zoologist B.S. Meadows carried out a study into a naturalised population of guppies that were living in a stretch of the River Lea which runs through Hackney in north-east London, where the temperature of the water had been raised to an ideal height for these tiny South American fishes. Ironically the population only survived because that portion of the river was so polluted that the only remaining native fish was the three spanned stickleback which did not pose a threat to the burgeoning population of tropical livebearers.

When he wrote his seminal book "The Naturalised Animals of the British Isles" in 1975, Sir Christopher Lever noted sadly, that although the gradual clean-up of the river could only be a good thing environmentally speaking, it would probably mean the end of the River Lea's population of guppies as larger, predatory species recolonised their old haunts, and munched away at the dwindling population of tropical interlopers. History doesn't relate whether Lever's dismal predictions were right or not, because around about 1990 the old coal powered power station was closed down for good, and whether or not there were any surviving guppies in the river, they would certainly have died out when the water temperature returned to normal. The other population of British guppies has an even more interesting – and tragic – history. In 1963, a pet shop in Lancashire went out of business and the proprietor heartlessly threw all his stock of tropical fish into the St Helen's Canal, where they would undoubtedly have died very quickly had a 400 yard stretch of the waterway not been heated to an ideal temperature for tropical fish by the

Poecilia reticulata
Population Trinidad, Caroni
Swamp
(N 10° 32.013'; W 61° 26.797')
Pair (above male, below female)
Photo: EHL 2013-01413

discharge from the nearby Pilkington Brothers Glass Factory. According to Lever, a viable breeding population of guppies and also of Red Bellied Tilapia (*Tilapia zillii*) was soon established along with non breeding populations of angel fish, mollies and an un-named species of tropical catfish. According to Leslie Bromilow, the secretary of the St Helen`s Angling Association, this stretch of canal became known as "The Hotties" and the tropical intruders survived happily there for many years. However, this story too has an unhappy ending. Leslie told us wryly: "unfortunately about ten years ago they switched the pumps off and the water cooled down and all the fish died, but there is still a thriving population of Red Eared Terrapins". We spoke to the press office at Pilkington Brothers who confirmed Leslie`s story, and although they bemoaned the fact that they had destroyed the country`s only surviving wild population of tropical fish, they were not at all impressed by our suggestion that the least that they could do was to release thirty quids worth of guppies back into the water and start again.

It does not take too much of a leap of the imagination to construct a scenario where something similar could happen with *Lampides boeticus*. The species feeds on a large selection of plants from the pea family, and with the proliferation of 'All Weather' attractions for both tourists and locals one can imagine such an attraction providing an environment where these beautiful little butterflies could 'come indoors, of its own volition in order to spend the winter months'.

Historically a few of these butterflies turn up in the UK each year but in 2013 an enormous number – estimated by some to be in excess of 10,000, turned up in the UK and bred extensively. We covered this in some detail a few issues ago. Last year, and this, butterfly enthusiasts have hope for a repeat invasion.

Sadly, this has not happened, although this year there have been several reports of successful breeding along the South coast, particularly in Sussex and Kent.

In 2014 a small colony of the European map butterfly (*Araschnia levana*) was found near Swanage in Dorset. It appears that a few specimens survived the winter but there is no evidence that the colony has continued for a second brood.

The map is unusual in that its two annual broods look very different. The summer brood are black with white markings, looking like a miniature version of the white admiral.

It has also been a disappointing year for sightings of the European race of the swallowtail. In 2014 they bred in a number of locations in the Southern counties but in 2015 there were only two records: one of the Scilly Isles, two in Sussex, and one in Devon. This, of course, does not include reports of the British race of this species which is confined to East Anglia.

The most exciting butterfly news of recent years took place in 2014 when a massive population explosion in the yellow legged or scarce tortoiseshell (*Nymphalis xanthomelas*) in Eastern and Central Europe caused the species to expand its range westward into Scandinavia and the UK; it has become naturalised in Sweden, and there were hopes that the same would happen in the UK.

However there have only been fourteen reports in 2015 from 9 different locations, which is – it is hypothesised – far fewer than would be needed for this species to have a realistic chance of becoming a UK resident.

The data and sighting reports in this article are taken from www.bugalert.net, and we are most grateful to Adrian Riley FRES., for all his hard work in maintaining the Bug Alert sight throughout the year.

SOCIAL NETWORKING CONUNDRA

Old hippies like me were vocal in our support for the concept of the 'Global Village' as espoused by people like Marshall McLuhan. The internet promised so much in this direction, and in many ways the 'Global Village' that we all wanted has arrived.

But I live in a real village, and it is not as idyllic as many people would like to think, sadly, the 'Global Village' has many of the same problems.

Let's talk about Facebook.

In many ways this social media website which was only launched eleven years ago by some roommates from Harvard, has changed the world more then anything else in the 21st century. Although some people don't have a Facebook account they are few and far between, and it is an incomparable boon to researchers like me to be able to type in somebody's name into a Facebook search engine and within seconds be in a position where one can write to the vast majority of people that one would wish to write to. There are Facebook groups for every conceivable interest, and I am proud to say that the CFZ Facebook group, which is moderated by Graham Inglis and Alex Clifford, has over a thousand members, and is growing every day.

But there is a downside this brave new world of social media means that everybody has a voice. But that's good isn't it? I can here you whisper in my virtual ear, isn't that what you wanted all along?

Well, yes. But the downside of everybody having a voice is – ironically – that everybody has a voice.

I am not going to even start on the social problems which come out of every teenager being able to shoot their mouth off about their feelings on their family and classmates without any realistic censorship. That is outside the remit of this magazine, which is – after all – broadly dedicated to cryptozoology and allied disciplines', but unfortunately this new found freedom for everybody to be able to say what they want about any person and any subject in packs upon our world (and by 'our', I mean those of us in the Fortean Zoological community).

Find us on: facebook.®

As I am sure that every reader of this magazine is aware, there are no university courses in cryptozoology. You cannot study for a recognised degree in the subject. There are college course in cryptozoology, mostly in America, but, and may I stress here that I am not belittling the content of such courses; I seriously doubt whether any of them come with any academic cachet worth speaking about.

As a result of this, anyone can call themselves a cryptozoologist. I know one bloke who was kicked out of school with four not very good O'Levels, and a redundant qualification in nursing the mentally handicapped who did just that. And in case you haven't figured it out, that person is me.

I have always tried to work within the scientific method basing my enquiries as far as possible on empirical evidence subject to specific principles of reasoning.

The CFZ has never been an organisation of Orwellian thought police. It has always been a broad church. Possibly too broad for some people's liking. However, we have always prided ourselves on adhering to a degree of internal logic.

Back in those halcyon pre-social media days, if we disagreed with someone, it was usually sorted out in an exchange of letters, to most people's satisfaction. Now, with anybody free to say anything we have a much more potentially volatile landscape.

There are all sorts of people doing all sorts of things broadly with a Fortean Zoological remit. Some of them are following practices in line with (or broadly so) what the CFZ promote. But others are most definitely not.

What makes it even more complex to deal with, is that occasionally some of the people in this latter category are broadly or tangentially linked with us.

Some of them proclaim there allegiance to the CFZ in large letters on their Facebook pages.

Now, I am a coward. I have always been a coward to a greater or lesser extent but now, at the age of 56, chronically ill, and with a wife, family and an ever growing tribe of extended members of my household to consider, I truly do not have the stomach for a ruckus anymore.

Over the last few years there have been a number of occasions when the CFZ (in general) and me (in particular) have been embroiled in long-standing, fruitless, and ultimately painful online disputes, which then got out of hand, achieving nothing apart from upset. I am too old and too tired to deal with this sort of crap anymore. This means that when somebody emails me complaining about somebody else, and belittling their beliefs, activities, or modus operandi, I usually try to ignore it.

The sad knock on effect of everybody having a voice is, as I have said, that everybody has a voice. And, nobody, is going to agree with everybody else.

At any one time there are bound to be quasi cryptozoological investigations going on somewhere in the world, with which I personally do not agree. Usually, when this happens somebody writes to me to complain. I would like to say here and now that, except in the most extreme of cases, I shall not be acting on this information. It is the role of the CFZ to lead by example rather than become involved in interminable online

disputes.

This brings us on to another, similar subject. I receive regular complaints from CFZ members, and indeed members of the public, about the behaviour of a surprisingly wide range of CFZ members, and affiliated persons, again on Facebook.

I would like to say once and for all that I am going to take no action just because a CFZ member allies themselves publically with a religion, sexuality, or political party with which somebody else does not approve.

In the last year I have had complaints that so and so was an atheist, that so and so had pictures of pouting girls in bikinis on their page, that another person was being vocal about his support for the Hunt Saboteurs, and that one person had posted Facebook

links to a perfectly legal political party.

Please note that unless somebody is promoting child abuse, animal abuse, blood sports (what may or may not be illegal but which are against the CFZ ethos) or a handful other examples of what are these days called hate crimes, I shall take no action whatsoever.

The vast majority of people who are banned from the CFZ Facebook groups are banned for abusive behaviour, posting pornography or trying to sell knock-off designer sunglasses.

Please forgive me for including such a long diatribe in this, the fourth magazine of the year, but it is something that needed to be said.

The last of the big red click beetles - a slightly optimistic tale of Danish cryptoentomology

by Lars Thomas

Once upon a time there was a large and colourful click beetle called *Ampedus praeustus*. Though it was sporting elytra in the most magnificent lacquer-red colour, it was by and large a peaceful and rather elusive creature like so many of its relatives. It preferred to spend its life munching away on old crumbling timber, or actually inside old crumbling timber like a large colourful termite. But like so many other insects and other animals that like old, rotting, crumbling, spongelike wood, it was not a fan of progress. The modern times did not sit well in click beetle land as it were, so although it was once fairly common and widespread in Denmark, it started disappearing more than 100 years ago. In the 1950s it had become a kind of holy grail for collectors, and the number of sightings dwindled to almost nothing, until it was actually considered extinct in Denmark in the 1980s.

The story could very well have ended here, but then things started to go all strange and cryptozoological. Stories started to emerge about people in a small part of Copenhagen harbour occasionally seeing large red beetles looking like nothing they had ever seen before – and with their strangely slim and elongated bodies, click beetles do look rather weird and unbeetle like.

Today that part of Copenhagen is sort of

the rundown backend of the harbour, with a mixture of small artists' workshops, slightly battered houseboats, the odd rundown hippie hideout and a bit of light industry thrown in for good measure.

But before that it was a major naval base for at least 500 years, which basically means there is a lot of old timber out there; old mooring posts, remnants of old bridges and dock walls, ruined buildings and bits and pieces of old wooden ships.

This all piqued the interest of now retired Danish entomologisk Dr. Ole Martin, who just happens to be an expert in click beetles. He arranged for several searches in the area, some of which had rather weird consequences.

When he started looking for *Ampedus praeustus,* part of the harbour was still an active naval base, as well as the HQ of Danish Intelligence and several other secretive organisations, so he had to search through old rotting timbers under the watchful and rather bemused eyes of a couple of heavily armed guards. In the late 1980s he did manage to find a couple of individuals of the beetle, but after that nothing.

A few years later the navy finally left the area, making further searches a bit less complicated, but no traces of the beetles were found. Although the odd story of people meeting red beetles kept surfacing.

And then we fast forward about 25 years, to the autumn of 2014, where Dr. Martin and a couple of friends made a last effort to find the beetle and ascertain the exact nature – if any – of its whereabouts.

Lo and behold – the beetles were still

there, although not at the former naval base. It turned out they were only to be found in about 50 pieces of old timbers along 20 metres of the edge of one of the canals in the area.

All of the timbers were located at the edge of a piece of undeveloped property, so it was clear from the outset, that if left to their own devices, it would only be a matter of time before they truly disappeared. *Ampedus praeustus* can still be found in Sweden and Germany, in limited numbers, so although its spreading potential is not especially good, it wasn't entirely impossible for it to turn up again. But nobody really believed that.

So, in the early summer of 2015, an agreement was reached with an institutions in the area to move the timbers there for the safekeeping of the beetles. To secure the privacy of the animals, I shall not name the institution nor the exact placing of their hometimbers.

The move was duly done on the 8th of July, and all went well. It even brought an added bonus, as the timbers also turned out to be the home of two more extremely rare beetles. Not bad for a morning's work.

Everybody involved is keenly aware that this does not solve the beetles' problem in the long run, but at least it buys everybody a little more time, as the researchers think the move has given the small beetle population another 30-50 years, and before they are gone, somebody may just have had a brilliant idea on how to save them on a permanent basis.

THE SERPENT DRAGONS OF ANCIENT GREECE

The dragons of ancient Greece were unlike most others in European tradition. Rather than the bat winged, quadrupedal creatures familiar elsewhere, Greek dragons were vast, limbless serpents usually furnished with a crest or fin on the head. Their breath was poison gas or venom rather than fire and they also killed by constriction. These Grecian mega-serpents tended to be blue in colour rather than the greens and reds that predominate elsewhere. They were not unlike the Germanic wurm, the Norse orm and the worms of Britain. They were foils for gods like Apollo and Zeus, and heroes like Jason and Cadmus.

However, paradoxically the Greek dragons were also said to be good spirits of life and regeneration. Agathos Daimon took the form of a crested serpent and was seen as an intermediary between mortals and gods. Emperor Noro, in the vanity of his madness claimed to be the new Agathos

Richard Freeman

Daimon and had coins minted declaring it! The precursor to Zeus was a giant bearded serpent called Zeus Meilichios. It is possible that this represents a far, far earlier god.

Snakes inhabited temples as sacred beasts and were tended to by priestesses. The Thesmophoria was a sowing festival performed by women, during which pigs were fed to serpents sacred to the corn goddess Demeter. It should be noted that no known snake of this region is large enough to swallow a pig.

Snakes were associated also with all things underground and were thought of as the vehicles of offended ghosts. Dead kinsmen, thought of as halfway between mortal and divine were said to take the form of snakes and dead heroes in the form of snakes were venerated across Greece in ancient times. Asklepios, who became the god of healing was one such hero. It was thought that the spines of decaying dead people would transform into snakes. Such ideas of the dead generating life in their decay were common in the ancient world.

What were these giant snakes of ancient Greece? There may be some clues in the artwork of the time. A krater (a vessel used for mixing wine and water) dating from 330 BC and held at the Musee du Louvre depicts hero Cadmus slaying a dragon. It is depicted with a bright red crest beard. It also has an unusually large, round eye. Another vessel held in the same collection depicts Cadmus struggling with a crested snake

The former of these in particular resembles the oarfish (*Regalecus glesne*). These are elongate fish with bluish silver bodies and striking, blood red crests (below). They can reach lengths of over 36 feet with unconfirmed reports suggesting that some individuals may reach 56 feet long. They have often been touted as an explanation for sea serpents by lazy sceptics. However

only a tiny percentage of sea serpent reports resemble this fish. Most involved dark bodied, vertically flexing animals with long necks held up out of the water or short necked creatures displaying a row of humps. None of these features point to the oarfish. The oarfish may not make a good sea serpent but it may have influenced Greek dragon lore.

Oarfish live at depths of around 3,300 feet and feed on crustaceans such as shrimp. They have noticeably large eyes. Despite their impressive size they are totally harmless to humans. When encountered by people at the surface they are usually in their death throes.

Imagine a fisherman seeing one of these beasts in the seas of ancient Greece or threshing about in the shallows. He would have no way of knowing that such an alarming beast was not a deadly serpent sent by the gods. Oarfish washing ashore have been linked to earthquakes. Dragons all over the world are also associated with earth tremors. In Japanese folklore Ryūjin is the dragon god of the sea, and the oarfish is its messenger. The message brought by dead oarfish is impending earthquakes and that appeared to be the case in 2011 when 20 oarfish came up on

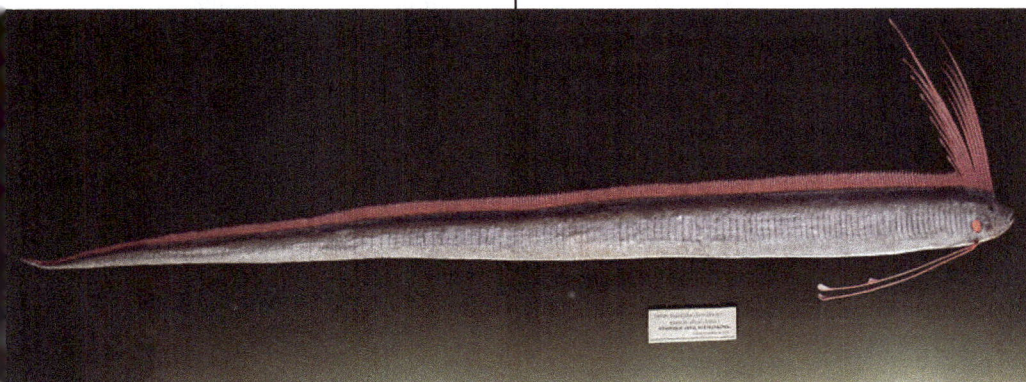

beaches in the area where the Tohoku earthquake and tsunami caused the most damage. Earthquakes release large quantities of carbon monoxide that can affect large deep sea creatures like the oarfish. The small fissures that precede major earthquakes could leak enough of the gas to make the fish sick and beach themselves before dying. Another possibility put forth by Rachel Grant, a lecturer in animal biology at Anglia Ruskin University in Cambridge, is electricity.

"It's theoretically possible because when an earthquake occurs there can be a build-up of pressure in the rocks which can lead to electrostatic charges that cause electrically-charged ions to be released into the water. This can lead to the formation of hydrogen peroxide, which is a toxic compound. The charged ions can also oxidize organic matter which could either kill the fish or force them to leave the deep ocean and rise to the surface".

We can readily see how the oarfish could have added to the legends of serpent dragons in Greece but there are other creatures behind the myths as well. The god Apollo slew a dragon called python at Delphi were it presided over the oracle. The dragon's name was later used to name a genus of large constricting snakes from the old world tropics.

Another piece of pottery held in a private collection, and dating from 350-340 BC, shows the dragon Lardon who guarded the golden apples in the Garden of the Hesperides. It is being fed by a priestess as it winds about a tree. The creature is transparently a python and from the detail in the depiction it is likely that the artist painted it from life. Its patternation and head structure most closely resembles the reticulated python *(Python reticulatus)*. This species can exceed 30 feet and is found in the tropics of South East Asia. This may seem a long way from Greece but the ancient Greeks were travellers and explorers who penetrated into both Africa and Asia. They would have been familiar with African rock pythons *(Python sebae),* the Indian python *(Python molurus)*, and the Burmese python *(Python bivittatus)*. The reticulated python comes from further east but it is not inconceivable that specimens may have made their way to Greece via traders. Such massive snakes would not have been transported as adults but as small youngsters. Kept in heated temples and fed and watered they could have reached huge sizes. Unlike the oarfish, a big python could easily kill and eat a man. Reticulated pythons and African rock pythons have been recorded doing just that.

The average person in ancient Greece would never have seen a snake much over five feet long. Imagine then, seeing a 20-30-foot snake in a temple. These would have been quite big enough to eat a pig, or a person. Kept well fed they would have been fairly docile and used to handling.

Intriguingly there is another possibility for the origin of the serpent dragon in Greece. One of the most famous Greek heroes is Jason. In legend Jason and the Argonauts sailed to the kingdom of Colchis in search of the golden fleece that was guarded by a dragon that never slept (note here that snakes have no eyelids). Colchis was a kingdom situated where the country of Georgia is today. In ancient times sheep's fleeces were used to filter gold in the area. This may be the origin of the Golden Fleece story. Greeks

traded in Colchis and the story of Jason and his epic journey may have had its genesis here. From Georgia and the Caucasus come reports of an unknown species of gigantic snake 20 to 33 metres long. These are reported as far east as Central Asia. Anatoly Serendenko, the Ukrainian archeologist who took part in the CFZ's 2008 almasty expedition to the Russian Caucasus had seen a 23-foot specimen in a cave. His father had seen another in a swamp in Kazakhstan. As the huge snake reared up he mistook it for a man standing up from a distance. If these monster snakes do indeed exist maybe one of them was behind the legend of the dragon that guarded the Golden Fleece and perhaps some of them were taken back to Greece to become sacred temple dragons.

Old Rome too was said to be home to monster snakes. The creatures were so vast they were said to feed on cows (*boves*) and were hence called boas from which we derive the modern name 'boa' for neo-tropical constrictors.

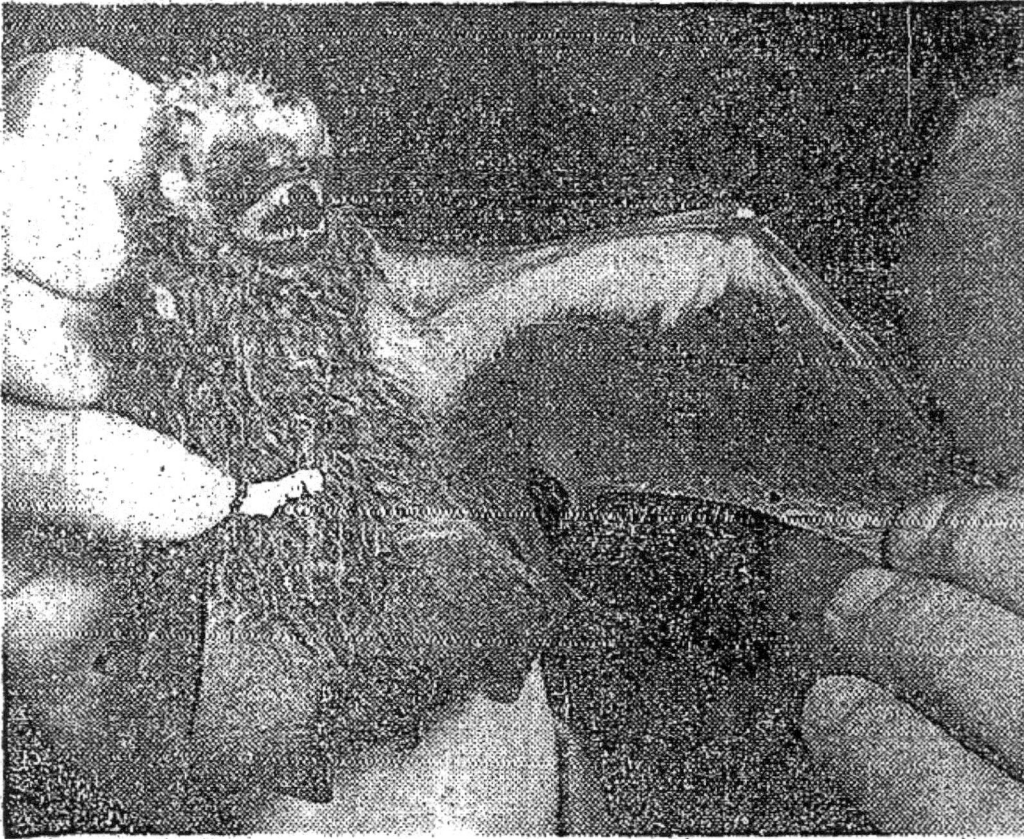

What was a "Vampire Bat" doing in Minnesota in 1955?

By Richard Muirhead

I recently came across this article in the *Winona Daily News* of October 8th 1955, a Minnesota newspaper; apparently concerning the surprising news of the capture near Dakota, Minnesota in 1955 of a silver-haired bat - a species that was then rare in the state, though it was initially thought to be a vampire bat, whose current and 1950s distribution stops around about the Mexico-U.S. border. The colour of this mystery Minnesota bat, however, doesn`t fit in with the silver-haired bat; it was described in the newspaper as being brown and white.

However brown and white does fit in with the common vampire bat. However neither the silver-haired nor the common vampire bat has a total length of 15 inches (= 381mm) which is the length of the Dakota cryptid-bat.

Here's The "Creature" brought to The Daily News office this morning by Mrs. Harry Golish and her daughter, Mrs. Roland Campbell, both of Dakota, who found it at a Dakota orchard Friday afternoon. Tentatively identified as a vampire, the animal has a wingspread of 15 inches, a brown and white furry body and sharp, white teeth, which were slashing at the cameraman at the moment this picture was taken. The hands in the picture are those of Dr. M. H. Doner, entomologist for the J. R. Watkins Co., who was called in to help identify the creature. Total length of the animal was about five inches. (Daily News photo)

●　　●　　●

By TOM BERGHS
Daily News Staff Writer

We can take the fish flies and the periodic onslaughts of crickets—if it becomes an utter impossibility to avoid them. . .

But there's just something d o w n r i g h t uncompromising about a vampire—especially if it be one with a wingspan of well over a foot. . . one that glares and hisses at you, as razor-edge teeth snap with the determination of a man-eater.

It's jolting to be sitting peacefully at 8 a.m. and suddenly have someone plop an animal in front of you that looks as though it would relish taking you apart limb from limb.

But that's exactly what a Dakota woman and her daughter did at The Daily News office this morning.

Mrs. Harry Golish and her daughter, Mrs. Roland Campbell, upended the newsroom early today when they dropped around with the vampire they'd captured at their home Friday afternoon.

No foolin!

They had with them a bottle . . . which bore within its confines a creature of the most ugly disposition we've ever seen.

The thing snarls at you and gives you a beady-eyed glare from beneath shaggy, tannish-yellow brows.

It had the face of a mean tomcat, the body of a furry lizard, the wings of a bat and a jawful of white, gleaming teeth.

●　　●　　●

Mrs. Golish and her husband found the creature about eight feet up in a tree at the South Wind Orchards, Dakota, where Golish is employed, about 4 p.m. Friday.

Golish, his wife reported, had ventured forth with pliers to remove the animal—kicking and screeching—from its perch.

Just about then Golish turned squeamish—and who wouldn't —so the couple hustled their daughter, Mrs. Campbell, to the scene.

The animal has a brown and white furry body, with a general appearance that bests anything Hollywood terrorists have managed to come up with.

Mrs. Golish revealed that the animal—in its frantic attempts to escape capture Friday—had cleanly bitten off the end of a one-inch stick with one slash of its vicious-looking teeth.

After a search through the encyclopedia failed to result in identification, Mrs. Golish had about made up her mind to return home and speciously avoid the area where the vampire had been found.

At the last minute, however, Hymes uncovered a musty publication concerning Minnesota mammals, which provided the answer.

The creature was identified as a silver-haired bat—rare in Minnesota. The biological name given was Lasionycteris Noctivagans.

It still looked like a vampire.

Mrs. Golish said they were going to take the animal home and mount it, but would prefer "never to find another one."

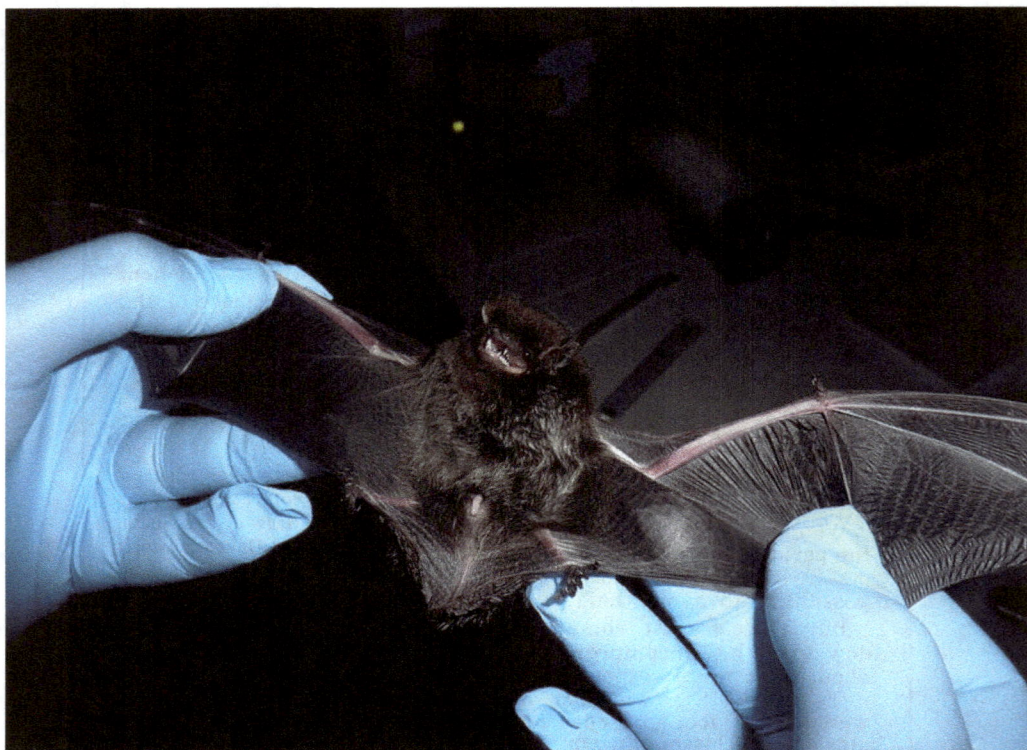

Male silver-haired bat captured in the Ozark National Scenic Riverways in 2010. (Larisa Bishop-Boros)

Common vampire bat, *Desmodus rotundus*, young male, captured near Lamanai, Belize (Gerry Carter)

The silver-haired bat reaches about 100 mm (= 3.94 inches) whilst the common vampire reaches 180mm (= 7 inches.) If it was a vampire bat it was further north than its usual range.

According to Wikipedia: Common Vampire Bat: "The common vampire bat is found in parts of Mexico, Central America, and South America.

They can be found as far north as 280 kilometres (170 mi) south of the Mexico–United States border. Fossils of this species have been found in Florida and states bordering Mexico. The common vampire is the most common bat species in southeastern Brazil. The southern extent of its range is Uruguay, northern Argentina, and central Chile. In the West Indies, the bat is only found on Trinidad. It prefers warm and humid climates, and uses tropical and subtropical woodlands and open grasslands for foraging. Bats roost in trees, caves, abandoned buildings, old wells, and mines. Vampire bats will roost with nine other bat species, and tend to be the most dominant at roosting sites.

They occupy the darkest and highest places in the roosts; when they leave, other bat species move in to take over these vacated spots."

The largest bats in the world are fruit bats and there is no way in which this bat was a fruit bat. This particular U.S. mystery bat isn't mentioned in Karl Shuker's *A Belfry of Crypto-Bats in Fortean Studies* volume 1.

And just in case you thought the world couldn't get any stranger than a large Minnesota mystery bat, there was also, as far back as January 1st 1909, according to the *Urbana Daily Courier*, an Illinois paper, a sea bat, which I think was some kind of ray.

THE SEA BAT.

Specimen of a Fish That Is Both Queer and Rare.

One of the rarest specimens of the fish kingdom known to waters contiguous to the North Carolina coast was captured in a seine at Masonboro sound by William Hewlett, a fisherman, says the Wilmington Dispatch. The fish, which was brought to the city, is what is called "the sea bat," and it is a perfect reproduction of a leather wing bat on a large scale. The fish is about fifteen inches long and about thirty inches across the back.

Strange to state, it had a thin, threadlike tail about fifteen inches in length, and on each side of the rear appendage were two perfectly formed gloved feet, with a smaller dimension having the exact appearance of a thumb with the other part of the hand mittened. The mouth of the strange specimen was about five inches across, and on each side of the mouth or the underside of the body there were five "strainers," or holes, through which the fish is said to rid itself of refuse products resulting from the forage it picks up at the bottom of the sea. The top of the fish was a dark slate color, and the under part of the body was white.

One old negro fisherman more than seventy years old declared that this was only the second specimen of the sea bat he had ever seen in his long experience as a fisherman. The specimen, which had a truly uncanny appearance, will probably be sent to the state museum at Raleigh.

OBITUARY

On November 6, I wrote an article for the Mysterious Universe website which began as follows:

'Back in the summer of 2003, I met – for the first time – a man named Rob Riggs. Like me, Rob had a fascination for the Bigfoot mystery. To the extent that he wrote a full-length book on the

ROB RIGGS REMEMBERED BY NICK REDFERN

subject. Its title, *In the Big Thicket: On the Trail of the Wild Man*. On the day in question, Rob proved to be very generous: he drove me through and around the Big Thicket, showing me the various Bigfoot "hot-spots" and places of historical significance.

'As I wasn't living too far away from him at the time (in Nederland, Texas), Rob and I stayed in touch and met up on a fairly regular basis. Then, in 2005, Rob organized the "Texas Ghost Lights Conference," which was held in the city of Austin, Texas. It was followed by a road-trip to – and a night-time vigil within – the Big Thicket. Unfortunately, I have just heard the very bad news that Rob has passed away.'

Altogether, I probably met Rob seven or eight times – on several occasions to explore the aforementioned Big Thicket and the rest at Texas-based cryptozoology-themed conferences.

I last saw Rob in October 2014, when I spoke at Craig Woolheater's annual Texas Bigfoot Conference, which was held in the town of Jefferson. I knew Rob had been sick, and I could tell that he wasn't feeling well at the event. But, I didn't know how ill he actually was – until I got the news of his passing.

I was particularly intrigued by Rob's work, as he didn't ignore the weirder side of Bigfoot and so-called "wild man"-type reports. Indeed, Rob spent a great deal of time addressing the matter of connections between the Big Thicket hairy-things and the area's resident 'ghost lights,' described as being very much like the far more famous 'Marfa Lights' of Texas.

He also noted links between sightings of Bigfoot and so-called ABCs, or Alien Big Cats, in the Big Thicket. Typically of the large and black variety.

That Rob did not dismiss the potential connections between Bigfoot and other Fortean phenomena is something which led him to suggest that Bigfoot is more than just an unknown animal. Rob also told me of odd synchronicities he experienced while investigating Big Thicket-based encounters with the creatures.

It's a big tragedy that we won't hear any more from Rob, but he left behind a good, solid legacy of material, books, and dedicated research – none of which should be forgotten or ignored. RIP, mate.

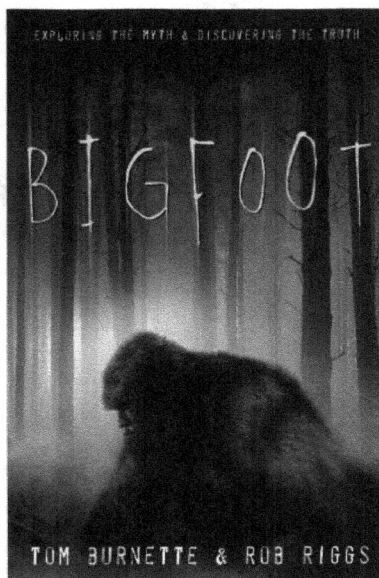

EXPLORING THE MYTH & DISCOVERING THE TRUTH

BIGFOOT

TOM BURNETTE & ROB RIGGS

DISCUSSION DOCUMENT: DUTIES FOR REGIONAL REPRESENTATIVES

The CFZ had had regional representatives for over twenty years now. Some of them have done remarkable things, some nothing at all, and some something in between. I originally intended my first wife to manage the list of regional reps, but as history shows, that never happened.

Ever since Alison and I split up I have been intending to ask someone else to take over the job, and finally a few months ago I got around to it. Ronan Coghlan has agreed to take over the onerous task, and has come up with a list of suggested roles

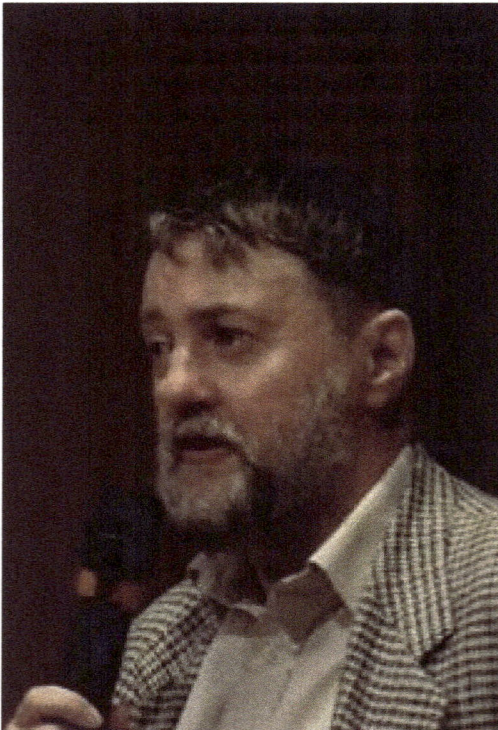

for regional representatives, which I post here for public discussion.

[1] In the event of a reported sighting of a mystery animal in the representative's area, all possible data should be gathered and forwarded to CFZ. Likewise, news of further developments should be sent on as they occur.

[2] Representatives should try to discover if there were any sightings or other anomalous events in their areas in the past, but should only send on stories of UFOs or ghosts if they consider them important, as otherwise their is the danger of CFZ being swamped.

[3] Representatives should, if possible, look into local folklore to discover if stories of anomalous events in the area occur. Liaisons should be initiated with the Bird, Butterfly and Conservation Officer in their areas where possible. They should, in addition, try to gather an archive of Fortean zoological material from their local studies libraries.

[4] Representatives should initiate liaisons with groups dealing with anomalies and nature in the area, provided they consider them and their personnel suitable.

[5] Representatives should have the option of offering sales of books to local bookshops. However, some might find this distasteful and so this should not be regarded as an actual representative's duty.

Letters

The editor and his compadres welcome letters for publication on all subjects covered by this magazine. However, we would like to stress that neither this magazine, or the CFZ are responsible for opinions expressed, which are purely those of the letter writer.

HALF AND HALF

Hi Jon,

A gynandromorph is an organism that contains both male and female characteristics. The term gynandromorph, from Greek "gyne" female and "andro" male, is mainly used in the field of entomology - the scientific study of insects.

These characteristics can be seen in the butterfly below where both male and female characteristics can be seen physically because of the sexual dimorphism butterflies display. Cases of gynandromorphism have also been reported in crustaceans, especially lobsters, sometimes crabs and even in birds. A clear example in birds is the gynandromorphic zebra finch. These birds have lateralised brain structures in the face of a common steroid signal, providing strong evidence for for a non-hormonal primary sex mechanism regulating brain differentiation.

A gynandromorph can have bilateral asymmetry, one side female and one side male (as seen below) or they can be mosaic, a case in which the two sexes are not so clearly defined. Bilateral asymmetry arises very early in development, typically when the organism has between 8 and 64 cells, later the gynandromorph is mosiac.

The cause of this mutation is typically, but not always, an event in mitosis during early development. While the organism is only a few cells large, one of the dividing cells does not split its sex chromosomes typically. This leads to one of the two cells having sex chromosomes that cause male development and the other cell having chromosomes that cause female development.

For example, an XY cell undergoing mitosis duplicates its chromosomes, becoming XXYY. Usually this cell would divide into two XY cells, but in rare occasions the cell may divide into an X cell and an XYY cell. If this happens early in development, then a large portion of the cells are X and a large portion are XYY, Since X and XYY dictate different sexes, the organism has tissue that is male and tissue that is female.

In his autobiography, *Speak, Memory,* the writer and lepidopterist Vladimir Nabokov describes a gynandromorph butterfly, male on one side, female on the other that he caught in his youth on his family's Russian estate. Gynandromporphic butterflies such as this are unfortunately non-viable - note the kink in the abdomen (half male, half female) of this specimen.

In my 8 years as a professional entomologist at Stratford Upon Avon Butterfly Farm I have still not been fortunate enough to witness a gynandromorphic specimen in the flesh. This perhaps testifies to the extreme rarity of this bizarre mutation. About a month ago we sent about 500 south American pupae to the Vannes Butterfly Farm in France and a few days ago we received these photographs from them of a superb gyandromorphic Queen Swallowtail *Papilio androgeus.*

How I wish this individual had emerged with us as we've had only four gyandromorphic specimens successfully develop in the last thirty years as they are an extremely rare, nonviable mutation.

Hope you enjoy
Carl. Marshall

I AM A MANWOLF

Jon,

It may interest you to know that the next release of Ubuntu Linux is code-named "Wily Werewolf", and is really rather good. People generally prefer it to Microsoft Windows (although the usual orders are on the lines of "Make this PC dual-boot Windows and Ubuntu").

So, your chance is fast approaching (again) to be using a cryptozoological operating system!

And yes, tech support may very well be forthcoming.

Dan H.

> I am a great fan of Bill Griffith, featuring Zippy the Pinhead, and I subscribe to his website which sends me a Zippy cartoon each day. When, one day a few weeks ago, Zippy met Bigfoot, I wrote to him and asked him for permission to reprint it. He was kind enough to say yes, so here it is…
>
> **http://www.zippythepinhead.com/**

Zippy the Pinhead meets Bigfoot, reprinted with kind permission of Bill Griffith

THE WORLD'S WEIRDEST PUBLISHING GROUP

We publish a lot of books. Indeed, I think that we could quite easily claim to be the world's foremost publishers of books about Fortean Zoology and allied disciplines, and our Fortean Words imprint is doing a great job in producing books on other non-zoological esoterica. However, I feel that it would be unethical to review our own titles. So here, to end this edition of *Animals & Men*, is a brief look at the books we have put out since the last issue.

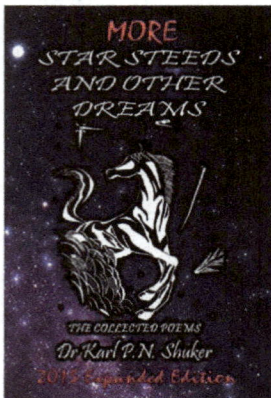

• **More Stars Steeds and Other Dreams: The Collected Poems. By Dr Karl P N Shuker**

Dr Karl Shuker has long been world-renowned for his numerous cryptozoological books, articles, and ShukerNature blog posts, but in 2009 his writings took a new and very unexpected, hitherto-unsuspected direction when CFZ Press published his first collection of poetry. Entitled Star Steeds and Other Dreams, its subjects were both diverting and exceedingly diverse - inviting its readers into Karl's very own, very personal world of star steeds and nightingales, childhood's end and silent farewells, realms of dreams and shadows, memory's mirror and ghosts from the past, Faerie worlds and flying horses, the voice of the winds and the music of the

spheres, roses and rainbows, airports, angels, balloons, butterflies, clowns, dragons, elves, fireworks, monasteries, poppies, Stonehenge, tattoos, UFOs, unicorns, and much else besides. Even Nessie, the Loch Ness monster, made an appearance. Much to Karl's surprise but delight, his poetry proved very popular, attracting interest and positive reviews both within and beyond his long-established readership audience, which encouraged him to continue writing poetry, experimenting with different styles and incorporating even more subjects within his verses.

Now, six years later, Star Steeds is back, but this time in a greatly-expanded 2015 edition, containing many new, additional poems and other lyrical writings.

So, without further ado, welcome to *More Star Steeds and Other Dreams* - enjoy the magic, the memories, and so much more that await your discovery amid the enchanted, and enchanting, realm that can only be found within the pages of this very different Karl Shuker book.

- **Sasquatch Down. By Michael Newton**

BODIES OF BIGFOOT: No less a personage than world-renowned paleoanthropologist Louis Leakey, when asked about Sasquatch, retorted, "Show me the bones."1 And indeed-under the International Code of Zoological Nomenclature and the International Code of Nomenclature for algae, fungi, and plants-each taxon (a group of one or more populations of a distinct organism) is based on a particular "type specimen." In order for Bigfoot-Sasquatch-Yeti to be scientifically classified, named, and (perhaps) legally protected, it must first be proven to exist. That basic requirement has spawned a fierce "kill/no kill" debate among researchers and monster hunters worldwide. The practical (some say "cold-hearted") pro-kill advocates insist that one specimen must be sacrificed for the good of all, and for the advancement of science. No-kill proponents suggest various alternative methods of proof ranging from photographs or videos-rejected sight-unseen by most skeptics-to plans for collecting flesh, blood or hair for DNA sequencing without harm to a living creature. So far, neither side has succeeded. Or, have they?

- **The Song of Panne (Being Mainly about Elephants). By Jonathan Downes**

A tired, crippled man in late middle age gets an unexpected visitor from his less then salubrious past. With him is something out of the pages of Greek mythology. The author finds himself deep in an arcane tangle of lies, half-truths and impossibilities. Against his better judgement, he has no option but to act. If this is a story, it's a bloody good one. If this is truth then the world you know is about to change forever.

- **Weird Wessex: A Tourist Guide to 100 Strange and Unusual Sights. By Andrew May and Paul Jackson**

At its height, the Saxon kingdom of Wessex sprawled across Southern England, encompassing Wiltshire, Hampshire, Dorset, Somerset and parts of Devon and Berkshire. Even before the Saxons arrived the area had a reputation as a weird place, with Stonehenge and its Druids, Glastonbury and the Holy Grail, the bizarre chalk figure of the Cerne Giant and the reputed location of King Arthur's Camelot.

In more recent times the tradition of weirdness has continued, with flying saucers sighted over Warminster, intricate Crop Circles popping up around Alton Barnes and hordes of spaced-out hippies converging on the mystical hubs of Glastonbury and Totnes. This book is a tourist guide with a difference.

It describes 100 of the weirdest sights in Wessex, ranging from world-famous places like Glastonbury and Stonehenge to hidden oddities that may even surprise the locals. Divided into ten thematic chapters, it is lavishly illustrated with over 200 full-colour photographs.

www.ingramcontent.com/pod-product-compliance
Lightning Source LLC
Chambersburg PA
CBHW050555280326
41933CB00011B/1860